This material is a gift

from the

LAKE COUNTY PUBLIC LIBRARY FOUNDATION

The
Creation/Evolution
Controversy

THE
CREATION/EVOLUTION
CONTROVERSY

A Battle for Cultural Power

Kary Doyle Smout

PRAEGER

**Westport, Connecticut
London**

Library of Congress Cataloging-in-Publication Data

Smout, Kary D.
  The creation/evolution controversy : a battle for cultural power /
Kary Doyle Smout.
    p. cm.
  Includes bibliographical references and index.
  ISBN 0–275–96262–8 (alk. paper)
  1. Creation. 2. Evolution—Religious aspects—Christianity.
3. Scopes, John Thomas—Trials, litigation, etc. 4. Evolution—
Study and teaching—Law and legislation—Arkansas. I. Title.
BS651.S59   1998
231.7'652—dc21   98–14927

British Library Cataloguing in Publication Data is available.

Library of Congress Catalog Card Number: 98–14927
ISBN: 0–275–96262–8

First published in 1998

Praeger Publishers, 88 Post Road West, Westport, CT 06881
An imprint of Greenwood Publishing Group, Inc.

Printed in the United States of America

The paper used in this book complies with the
Permanent Paper Standard issued by the National
Information Standards Organization (Z39.48–1984).

10 9 8 7 6 5 4 3 2 1

Copyright Acknowledgments

The author and publisher are grateful to the following for granting permission to reprint
from their materials:

Excerpts from William Jennings Bryan, *Seven Questions in Dispute* (New York: Fleming H.
Revell, 1922), appear courtesy of Fleming H. Revell, a division of Baker Book House Company, Grand Rapids, Michigan.

**For Synthia,
Helpmeet**

Contents

Preface

This book began as an effort to understand a heated battle involving the meanings of a few key words. The words are *creation, evolution, science, religion,* and *truth,* as they have been used in the battle about whether creation or evolution should be taught in science classes in American public schools. However, just as a whole garment sometimes unravels if one tugs too long at a single thread, my own thread of investigation gradually lengthened until I saw that I was questioning and challenging some important assumptions about language itself that seem widely shared throughout American politics and culture. Let me briefly lay out the contours of this unraveling starting from its first loose stitch.

While studying English language and literature in college and graduate school, I became aware of the many competing accounts of language that have been developed this century in a wide variety of academic disciplines. Many scholars have seen this proliferation as a crucial turning of interest toward language itself; they have labeled it, variously, "the linguistic," "the interpretive," and "the rhetorical" turn. Especially intriguing to me were the claims made by those scholars commonly called deconstructionists or poststructuralists that language is not a stable namer of reality, what Richard Rorty has called a "mirror of nature," but a glass through which we see darkly, a shaper of the reality that we perceive. When I learned that contemporary poststructuralists have a view of language in some ways similar to that of the classical rhetoricians in ancient Greece and Rome, I eagerly turned to a study of rhetoric in hopes of developing an account of language that would synthesize what I had learned.

I found that whereas many Western thinkers since Socrates have thought of words as names for objects (like chairs) or ideals (like justice and reason), some classical rhetoricians and contemporary poststructuralists have thought of words instead as practical instruments used by groups of people to work out social arrangements and to achieve common goals. In order to work together, people must not only agree on the meanings of key words, but figure out how to apply those words to particular situations. Thus I began to see language as a great arena for human persuasion rather than formal demonstration, as a function not of abstract impersonal realities, but of concrete human communities.

After developing a rhetorical account of language, I started to search for an important problem plaguing a variety of disciplines on which I could test it. My goal was to give a convincing poststructuralist analysis of the language used in a major clash in American society. At first I considered abortion, in part because I noticed a striking pattern in this public debate: many, if not most, participants seem to assume that if they properly define the key term (*life* or *choice*), their opponents will have to agree with the position they ultimately take. I began to wonder why so many people in our culture would make this assumption about language. I thought next about tackling a recent debate within the English profession: perhaps literary theory, the canon, multiculturalism, or the goals of humanities education. These debates seem to follow the same basic pattern as the debate about abortion: participants attempt to resolve the disagreement simply by defining key terms; but their opponents regularly reject their choice of terms, their definitions, or both. I finally settled on the creation/evolution controversy. This debate has the advantage of a long and illustrious history, having started with Charles Darwin's *Origin of Species* in 1859 and continued intermittently through such spectacular episodes as the Scopes "Monkey" Trial to the present day. It involves a broad range of disciplines, including anthropology, biology, geology, history, law, paleontology, philosophy, religion, sociology, and theology. It is unusual enough to be interesting to a nonscientist yet can be understood without much scientific expertise; indeed, it has been argued almost exclusively in the public arena, where the issue of expertise itself is debated. Moreover, it focuses squarely on terminological issues, returning unceasingly to the proper definitions of such key terms as *creation*, *evolution*, *science*, and *religion*, with both sides insisting on the correctness of their own definitions and rejecting the definitions of their opponents.

When I began the study, I wanted to understand why evolutionists and creationists would fight so bitterly about the meanings of these key terms but never make any progress toward resolving their disagreement. I knew that although both sides had prevailed at various times, neither side had ever convinced the other. If definitions of terms worked the way that both sides assumed, why had they never resolved the conflict? I decided to focus my investigation on three key confrontations that were decisive and complex: the initial scientific debate about Darwin's *Origin of Species*, culminating in the 1860 Huxley/Wilberforce debate, the 1925 Scopes "Monkey" Trial, and the 1981 Arkansas Creation-Science Trial. In reading all available accounts of these episodes, I started to see a striking pattern: the initial debate always centered on a disagreement about the key terms, but later accounts portrayed the debate as a battle between obvious truth and obvious error, in which both sides consistently accused their opponents of ignoring the plain truth, as expressed in simple language, and instead using the language of rhetoric to deceive. Both sides assumed that the truth was obvious and that anyone should be able to see it; as a result, they could not account for their opponent's reactions as anything but obstinate blindness. However, the fact that their opponents could not see the plain truth proved that the truth was not so plain after all.

This book attempts to narrate this recurring story of mutual blindness from a poststructuralist perspective. It argues that both the creationists and the evolutionists misconceive of language as a simple mirror for reality instead of as a tool used to create and sustain various human communities. When these communities conflict in their conception and use of key words and meanings, they cannot call down from heaven a dictionary to resolve their disagreement; instead, they must find ways to persuade the culture as a whole to accept their definitions of terms and to enact their position as public policy. However, because our culture accepts an oversimplified view of language, neither side sees its work as cultural persuasion. Instead, they keep trying to demonstrate the truth. They view language as a naming operation rather than a complex tool that always, and unavoidably, reflects basic human values and beliefs.

What started out as an effort to understand a particular controversy thus became a study of how language is conceived and used in American culture. Instead of finding correct meanings for all the key terms, I discovered a linguistic competition between two worldviews, which I shall call, for simplicity, the humanist worldview and the fundamentalist worldview. Creationists and evolutionists overtly argue about the proper definition of *science*, but when the thread of their logic is pulled far enough, they turn out

to be arguing about whose worldview, whose conception of reality, will prevail—and in what political system. They both begin from significantly different assumptions, tell powerful opposing stories that give coherence to
their lives, and accept different basic values and beliefs that are invisible to
them and inscrutable to their opponents, but that structure their views of
everything else. Their battle is a battle about whose worldview shall prevail
in American culture. I believe that most contemporary battles in American
culture are, finally, similar arguments about whose values and beliefs shall
prevail. However, people on all sides argue for their worldviews not on the
basis of values and beliefs, but on the basis of the word *truth* itself. They attempt to specify what counts as truth in words which they claim (despite
enormous evidence to the contrary) have obvious meanings.

"We hold these truths to be self-evident" proclaimed Thomas Jefferson
in a seminal American statement. If the truths were self-evident, then why
did Jefferson need to proclaim them? I think he meant instead something
like "We believe" rather than "We hold," but he worried that *belief* might
not be a strong enough word in this sentence, especially in a nation largely
dedicated to erecting solid truth that could stand independent of the shifting sands of belief. The creation/evolution controversy is, finally, a debate
about the possibility of separating truth from belief, about the grounds of
liberal political philosophy and its culture of expertise. Both sides are ultimately arguing about how to see reality and how to organize our political
life, just as in similar battles between humanists and fundamentalists in
many nations around the world. As we debate the future of the United
States and what it considers its truths, I think we need a better account of
language. This book develops such an account and applies it to a heated battle. I hope it sheds light, not only on this battle, but on the shape of many
other cultural battles in our troubled and disputatious time.

For all their help with this book, I am grateful to Ron Butters, Stanley
Fish, George Gopen, Stanley Hauerwas, George Marsden, George Ray, and
Barbara Herrnstein Smith.

The Creation/Evolution Controversy

1

Introduction

In his 1867 book *Language and the Study of Language*, the American linguist William Dwight Whitney describes language acquisition as an interactive human process and then poses a problem of human communication that I shall take as the focus for this study. Whitney explains that children learn language, not just through their own efforts (or through what formalist linguists such as Noam Chomsky have characterized as the growth of grammar in the brain), but also through meticulous teaching by the adults who live with them.[1] Through this teaching, children learn not just how to speak correctly, but also how to use language to accomplish their goals.[2]

After explaining his insight that people learn words from other people, Whitney turns to the problem of disagreements about word meanings. He writes: "Not all who speak the same tongue attach the same meaning to the words they utter. We learn what words signify either by direct definition or by inference from the circumstances in which they are used. But no definition is or can be exact or complete; and we are always liable to draw wrong inferences" (20). A vital fact about the meanings of words is that people do not learn them by rote, as they memorize the multiplication tables or algebraic equations. Instead, they encounter new words in particular contexts, where they can either infer the meanings or ask for direct definitions, both processes that are fraught with the possibility of error. In both cases, the listener can still misunderstand the meanings of the words.

Whitney next uses a set of interesting metaphors to suggest why such misunderstandings can occur:

Words are not exact models of ideas; they are merely signs for ideas, at whose significance we arrive as best we can; and no mind can put itself into such immediate and intimate communion with another mind as to think and feel precisely with it. Sentences are not images of thoughts, reflected in a faultless *mirror*; nor even *photographs*, needing only to have the colour added: they are but imperfect and fragmentary *sketches*, giving just outlines enough to enable the sense before which they are set up to seize the view intended, and to fill it out to a complete picture; while yet, as regards the completeness of filling out the details of the work, and the finer shades of colouring, no two minds will produce pictures perfectly accordant with one another, nor will any precisely reproduce the original. (20–21, italics added)

Using incremental imagery of the mirror, the photograph, and the sketch (favorite images throughout Western history for discussions of representation), Whitney gradually displaces the link between a word and its meaning from a "faultless reflection" to an "imperfect and fragmentary sketch." He argues that this link cannot be precise because words are used by minds that do not precisely coincide. This imprecision is not caused by a flaw in language, but by unavoidable differences between people, who use the same tokens of language to express different thoughts, to produce different pictures. Word meanings are different because people are different.

In a final quotation, Whitney explains what can happen when people disagree:

[Not only words but] every part of language [is] liable to be affected by the personality of the speaker, and most of all, where matters of more subjective apprehension are concerned. The voluptuary, the passionate and brutal, the philosophic, and the sentimental, for instance, when they speak of *love* and of *hate*, mean by no means the same feelings. How pregnant with sacred meaning are *home*, *patriotism*, *faith* to some, while others utter or hear them with cool indifference! It is needless, however, to multiply examples. Not half the words in our familiar speech would be identically defined by any considerable number of those who employ them every day. Nay, who knows not that verbal disputes, discussions turning on the meanings of words, are the most frequent, bitter, and interminable of controversies? (21–22)

What Whitney has earlier hinted at as the possibility of semantic error and verbal disagreement suddenly erupts in this last sentence into a direct and heated conflict over the meanings of words. Especially with such everyday words as *patriotism* and *faith*—words that few people would "identically define"—people may become embroiled in controversies that are "frequent, bitter, and interminable." Whitney believes that such controversies are so common that he need not provide a single example; he assumes that the reader will accept his three superlative adjectives without question. Although he provides no proof, his thesis is supported, not only by my personal experiences, but by such recent battles in the American public arena as the Vietnam War, the Watergate scandal, abortion, affirmative action, and the censorship of art; and in college and university English departments by battles over multiculturalism, the literary canon, and literary theory. Each of these conflicts involves a disagreement about basic terms, but none has been resolved through the specification of correct meanings. These particular instances certainly seem "frequent, bitter, and interminable." I shall call such conflicts over the meanings of words *terminology battles*.³ This book is an attempt to understand one such terminology battle, to trace its history, and to see what it suggests about similar battles that continue to recur in the United States.

TWO ACCOUNTS OF LANGUAGE

Some people might answer questions about the meanings of words by relying on what I shall call a *positivist* or a *philosophical account* of terminology battles. According to this account, such battles exist because some people are not careful enough about their words, especially words that represent abstractions such as Whitney's *patriotism* or *faith*. This lack of care sometimes prevents them from understanding one another and thus from agreeing on the best solution to a common problem. In order to solve such a problem, this account suggests, the antagonists must define their terms carefully, perhaps by relying on an authoritative dictionary, and then stick to those definitions. Each word needs to be attached to the thing it represents; it needs to be tied down, kept from slipping; made tighter and sharper, more exact, and more precise and rigorous: that would solve their language problem, at least. This account suggests that if everyone would be more careful and precise, there would be no word-meaning problems to resolve.

I call this account *positivist* because it assumes a positive link between a word and the thing it represents in terms of a conception that has dominated much Western thinking about the word-meaning link for over 2,000 years. I call it *philosophical*, perhaps unfairly, because such a conception was at least implied by Socrates when he asked his interlocutors for a precise definition of terms, and it has been advanced periodically by philosophers in the intervening years. Indeed, it was made into an overt goal of some Enlightenment philosophers. Bishop Thomas Sprat, a historian of the British Royal Society, indicated this goal as follows:

to reject all the amplifications, digressions, and swellings of style: to return back to the primitive purity, and shortness, when men deliver'd so many *things*, almost in an equal number of *words*. They [the Royal Society] have exacted from all their members, a close, naked, natural way of speaking; positive expressions, clear senses; a native easiness: bringing all things as near the Mathematical plainness, as they can: and preferring the language of Artizans, Countrymen, and Merchants, before that, of Wits, or Scholars.[4]

In the twentieth century, this view is most widely associated with the logical positivists. Regularly problematized by other philosophers throughout history, it has also regularly reappeared in a variety of forms, often under a label like "the plain style." This view implies that disruptions in a word-meaning link create confusion that may escalate into a terminology battle. However, whenever a careful thinker gives a rigorous definition of a key term such as *justice*, he or she unmasks the sources of disagreement and presents the correct meaning so clearly that the language problem disappears. A positivist thus resolves a terminology battle by precisely specifying the relevant word meanings.

In Whitney's work on linguistics can be seen in miniature an opposing view of terminology battles, which I call a *rhetorical account*. (These opposing terms are also used to refer to the periodic conflict throughout Western history between rhetoric and philosophy.) In brief, some classical rhetoricians, such as Protagoras, Gorgias, and Isocrates, rejected Plato's conceptions of philosophy and saw language more practically.[5] From them I have learned to see language, not as a mirror for the world, but as a set of tools that can be used for all kinds of human purposes, as a creative instrument with which to solve human problems rather than as an imperfect namer of abstract and concrete realities. Within a rhetorical conception of language, word meanings cannot be specified positively—once and for all through word-thing correspondences or precise definitions. However, they can be specified rhetorically—with meanings and definitions that work well in a particular time and place for a particular group of people because of successful acts of persuasion. Such a rhetorical act creates what seems to be a positive fact about the meanings of words, a certainty that can itself be used as part of a persuasive strategy. When authority figures claim to be revealing correct word meanings, they are using a tactic of persuasion that has proven very successful in the Western tradition. They use this tactic to cause their own word meanings to prevail.

LANGUAGE AS A NET

If language does not give names to preexisting realities, then how does it work? This has been one of the major problems for twentieth-century language studies. From poststructuralist accounts of language I have learned to conceive of language, not as a chain linking a word to the thing it represents, but as a net in which a word takes shape fluidly, in relation to its surrounding neighbors. In his 1915 *Course in General Linguistics*, Ferdinand de Saussure contends: "The concepts [of language] are purely differential and defined not by their positive content but negatively by their relations with the other terms of the system. Their most precise characteristic is in being what the others are not."[6] Saussure compares a word to a particular piece in a chess game, which gains its value in the game, not from its intrinsic properties, but from its relations to the other pieces; a knight is a knight because it is not a king, queen, bishop, rook, or pawn.[7] To apply this point to the sounds of language, the word *cat* retains its identity by sounding different from *cab, can, mat*, and *bat* and by meaning something other than *dog* or *bird*. The individual strands in the net are held in place by other, nearby strands; they can be specified only by rigorously separating a particular strand from its nearest neighbors. According to this conception, one cannot make sense of words and their meanings individually, but only in terms of the dichotomies, or oppositions, in which they function.

In a rhetorical conception of language, these dichotomies have another crucial quality: they do not work as equal opposites, but as implicit hierarchies, which value one term and devalue its opposite, as in *good* versus *evil*, *light* versus *darkness*, and *presence* versus *absence*. First introduced by Jacques Derrida, this notion of hierarchies is explained most clearly by his translator, Barbara Johnson:

> Western thought, says Derrida, has always been structured in terms of dichotomies or polarities: good vs. evil, being vs. nothingness, presence vs. absence, truth vs. error, identity vs. difference. . . . These polar opposites do not, however, stand as independent and equal entities. The second term in each pair is considered the negative, corrupt, undesirable version of the first, a fall away from it. Hence, absence is the lack of presence, evil is the fall from good, error is a distortion of truth, etc. In other words, the two terms are not simply opposed in their meaning, but are arranged in a hierarchical order which gives the first word *priority*, in both the temporal and qualitative sense of the word.[8]

Because these dichotomies are hierarchical as well as descriptive, each word expresses human values and emotions even as it expresses concepts. One does not just perceive a reality and select its matching word; in the very choice of a word with which to describe a segment of reality, one reacts evaluatively to that reality, both in choosing that particular word as the best word and in dealing with what I shall call its *valence*, its overtones in its relevant dichotomy.[9]

An example may help clarify this point. In the past three decades, many feminists have attempted to specify the implicit evaluations of the male/female dichotomy, to show that *male* is regularly a positive term and *female*, a negative term. A good example of the valences of these terms (their implicit positivity or negativity) appears in the children's movie *Mary Poppins*. In one scene, the thoroughly chauvinistic father, Mr. Banks, decides to take his children on an outing to the bank where he works; he explains that he wants to give them a break from all the "sugary female thinking" they get at home. Within the associations created by this movie, his word *female* does not objectively describe a particular style of thought; instead, it actively devalues women, demeaning Banks's wife and house servants, who are listening as he speaks. Banks's use of the word *thinking* ironically suggests that real thinking never occurs in this house full of women, but occurs in abundance in the bank, which is full of men. His use of *sugary* links this type of thought to candy, thus associating it with the immaturity of children and distancing it from the maturity of adults, as well as suggesting decay through candy's effect on the teeth (although the movie reverses these associations later by having Banks himself learn that a "spoonful of sugar" helps him after he loses his job at the bank). According to a rhetorical account of word meanings, words like *female* always evaluate, as well as label, their referents. In this instance, the words in "sugary female thinking" do not simply label realities; they also convey values and feelings within complex linguistic nets.

A rhetorical account of word meanings thus crucially depends on identifying the relevant dichotomies for a particular use of a word and on examining the functions of these dichotomies, both conceptually and emotionally. Rather than thinking of a word and its meaning in isolation from other words, one needs to conceive of the key terms of a text as parts of intricate nets used by people to achieve their goals.

LANGUAGE AND HUMAN COMMUNITIES

Central to a rhetorical account of language is the notion of the individual speaker or writer as a member of a human community. Whitney imagined this community as comprised of the adults who teach a child to speak, but the notion has also become a major focus of recent poststructuralist language studies. Among the many poststructuralist writers who have attempted to reconceive of language as a function not of abstract and timeless realities, but of changing human communities are Stanley Fish, Michel Foucault, and Alasdair MacIntyre.

Fish is a well-known American literary critic who helped to introduce poststructuralism to the United States in the form of reader-response criticism of literary works. He wrote the first full book of reader-response criticism to be published in America, a 1967 study entitled *Surprised by Sin: The Reader in "Paradise Lost."*[10] In his subsequent studies of readers and interpretations, Fish began to develop a notion that seems closely related to the philosopher Ludwig Wittgenstein's conception of a language as a form of life. Fish introduced this notion as the "interpretive community" in his 1980 book *Is There a Text in This Class? The Authority of Interpretive Communities.*[11] Rejecting a conception of interpretation as a simple process of decoding words recorded on the page, Fish argued that readers learn to interpret literary works—and indeed, all language acts—not through some mechanical formula, but through their experiences as members of human communities. According to Fish, the meaning of a text is discovered, not just by assigning set meanings to all the words and then determining their syntactic relationships, but also by learning what it means to be a person in a particular situation who is using language to do a particular task. In short, a text can only be interpreted within a human context. In more recent works, Fish has applied this notion to other interpretive communities besides literary critics, most notably to lawyers and judges.[12] By comparing communities of legal professionals to communities of literary critics, Fish has continued to explore his central notion of the interpretive community as the key to human understanding. Fish contends that people function as members of various interpretive communities, in which they learn from other people how to assign meanings to words, texts, and events. Different communities will assign different meanings to the same texts as a result of their intricately woven linguistic nets.

Interpretive communities are formed by virtually any group of humans who try to make sense together. Members change and interact with each other, using language as a tool to define the group and to keep it working together. Besides the family mentioned by Whitney, one can find such communities in businesses, churches, towns, professions, even the academic disciplines themselves. I first encountered the notion of the academic disciplines as language practices with a history in the work of Michel Foucault, a French poststructuralist philosopher who has traced the development of language itself into an instrument of discipline and power. In his rich and complicated accounts, Foucault suggests that the academic disciplines have provided ways for many people in the last century to earn their living and to exercise the power of expertise over other people by mastering the arts of language as it is conceived and used by their particular disciplines. His best-known account of this process is the first volume of his monumental and unfinished study, *The History of Sexuality* (1980); in this book, he argues that discourses dealing with sexuality were developed only recently within the fields of psychology and medicine as a way to regulate the actions of patients by exercising linguistic and cultural control over them.[13] Especially striking in this book is Foucault's poststructuralist account of power as a structure of relations rather than as an object possessed by the powerful and used against the powerless; this account suggests that one exercises power whenever one attempts to influence the actions of another person by any means whatsoever (most often, by talking or writing to him or her). Foucault's accounts suggest that language is an instrument of human persuasion, through which power is exercised in a very concrete way. He focuses especially on the increasing power of the academic disciplines as influential human communities.

Another crucial writer who has influenced my thinking about human communities is Alasdair MacIntyre, an American philosopher who has applied a poststructuralist view of language to the concepts of reason and justice. In his 1988 book, *Whose Justice? Which Rationality?* MacIntyre argues that individuals learn to think and speak as members of communities who share a tradition; he believes that "rational enquiry [is] embodied in a tradition" rather than being a uniform and universal process for all people, as the Enlightenment thinkers held.[14] MacIntyre insists that Enlightenment accounts of universal human justice and reason can no longer be accepted because recent studies in many disciplines have shown that there are no rationalities independent of particular traditions, and thus that the words *reason* and *justice* mean different things for different individuals in different contexts. His argument implies that each tradition or community has its own account of what it means to be a rational and just community member. After tracing his thesis through four different philosophical traditions, MacIntyre concludes that there are multiple justices and multiple rationalities, all conceived by what Fish would call different interpretive communities. Each community defines these desirable concepts for its own members, who in turn define themselves in relation to these ideals and expectations and work to embody them.[15] These members use language as an instrument of persuasion in order to invite other members to join them in their particular search for knowledge and virtue.

BATTLING FOR CULTURAL POWER

By thinking of words, not as permanent names for abstract things, but as tools used to create and sustain human communities, one begins to see terminology battles differently. They are not simple efforts to specify correct definitions, but bids for power made by various individuals within competing communities, all of whom try to authorize their own definitions for key terms as a function of their own worldview, and thus to specify for the culture as a whole what those words ought to mean. Such controversies involve competing traditions of reason and justice, or more broadly, competing worldviews that are fighting for a recognized place in American culture.

From a rhetorical perspective, a terminology battle can thus be seen, not as a stubborn refusal to accept correct definitions of terms, but as a power struggle between competing communities. These communities try to convince other communities that their own word meanings make the best sense. The problem is that in a culture based on Enlightenment conceptions of a universal reasoning faculty in humans, people do not ask, "Best sense according to whom?" In effect, the terminology battle becomes a battle about worldviews. Those who win this battle attain the power to define the terms from within their own worldview for the culture as a whole.

The following chapters apply this rhetorical conception of word meanings to a particular conflict that involves a terminology battle: the recurring controversy about whether creation or evolution should be taught in science classes in American public schools. Each chapter looks at some of the crucial words, dichotomies, participants, interpretive communities, and institutions involved in a representative confrontation in this decades-long battle. My analysis illustrates in detail the ways in which opponents on both sides of this issue have used language to defend their definitions of terms within significantly different worldviews and to argue for their respective positions in order to influence public policy. Their efforts suggest concretely that word meanings are not positive facts, but rhetorical tools used by differing communities to compete for cultural power.

This controversy began with the publication of Darwin's *Origin of Species* in 1859 and has been refought many times in the last 150 years without any sign, then or now, of ultimate resolution. It crucially involves the meaning of several key terms, especially the term *science*. The next chapter summarizes the beginnings of this controversy by describing the battle about whether Darwin's *Origin* was itself a scientific work. Then it analyzes a famous debate between the evolutionist T. H. Huxley and the creationist Samuel Wilberforce, which occurred at Oxford in 1860. The third chapter focuses on William Jennings Bryan and the famous Scopes "Monkey" Trial of 1925. This first great episode of the controversy in the United States involved the attempt to outlaw the teaching of evolution in Tennessee public schools. The fourth chapter analyzes the 1981 Arkansas Creation-Science Trial. This recent battle had the opposite goal: to prevent creationism from appearing in public school science classes. These chapters collectively argue that the creation/evolution controversy involves a paradigmatic terminology battle between competing interpretive communities. The conclusion then develops the political implications of this battle; it suggests that in this controversy we are witnessing an ongoing debate between humanists and fundamentalists about the validity of liberal political philosophy and its reliance on the knowledge of experts to create a just and rational society through the law. Because American culture continues to hold a positivist conception of truth and a negative conception of rhetoric (as language used to deceive), the group temporarily wins that uses its own rhetoric to convince the nation that rhetoric was not used and that the truth presented itself. Let us see how this tactic has been used.

NOTES

1. Chomsky's clearest account of language acquisition is *Language and Problems of Knowledge: The Managua Lectures* (Cambridge: MIT Press, 1988). See especially pages 4, 35–36, and 149–151.

2. William Dwight Whitney, *Language and the Study of Language*, 5th ed. (New York: Charles Scribner's Sons, 1887), 11–15. Subsequent page references appear in parentheses in the text.

3. This term is adapted from Whitney's last sentence. The other key terms for this account, *word* and *meaning*, have been chosen only after careful consideration and because they are the ones most often used in terminology battles (as evidenced by Whitney): disputed sentences regularly take the form "X is Y" or "The word X means Y." These terms thus provide a concrete way of reenacting the latter statement, despite potential problems with their precision and rigor. For a discussion of these problems, see B. F. Skinner, *Verbal Behavior* (New York: Appleton-Century-Crofts, 1957), 7; and Barbara Herrnstein Smith, *On the Margins of Discourse: The Relation of Literature to Language* (Chicago: University of Chicago Press, 1978), 88–99.

4. Quoted in S. Michael Halloran and Merrill D. Whitburn, "Ciceronian Rhetoric and the Rise of Science: The Plain Style Reconsidered," in *The Rhetorical Tradition and Modern Writing*, ed. James J. Murphy (New York: Modern Language Association, 1982), 65–66.

5. Some excellent introductions to the history of rhetoric are Stanley Fish, "Rhetoric," in *Doing What Comes Naturally: Change, Rhetoric, and the Practice of Theory in Literary and Legal Studies* (Durham, N.C.: Duke University Press, 1989),

471–502; Edward P. J. Corbett, "A Survey of Rhetoric," in *Classical Rhetoric for the Modern Student* (New York: Oxford University Press, 1965), 535–568; Peter Dixon, *Rhetoric* (London: Methuen, 1971); and Thomas J. Conley, *Rhetoric in the Western Tradition* (New York: Longman, 1990).

6. Ferdinand de Saussure, *Course in General Linguistics*, ed. Charles Bally and Albert Sechehaye, trans. Wade Baskin (New York: McGraw-Hill, 1959), 117.

7. Ibid., 110.

8. Barbara Johnson, Introduction to Jacques Derrida's *Dissemination*, trans. Barbara Johnson (Chicago: University of Chicago Press, 1981), viii.

9. The notion of valence is similar to the common notion of the connotation of a word, but because a poststructuralist conception of the word-meaning link seems to me to empty the concept of denotation, I hesitate to keep *connotation* without its defining opposite.

Another possible label for this other aspect of a word besides its meaning is the term *emotive function*, as used by C. K. Ogden and I. A. Richards in *The Meaning of Meaning*, 8th ed. (New York: Harcourt Brace, 1956). However, their opposite for this term is *referential function*, a notion quite close to "the thing behind the word." Neither *connotation* nor *emotive function* imply anything about words as instruments of value, a use that seems to me to be the crucial issue.

Ogden and Richards spend five pages attacking the notion of connotation itself (186–190). Then they specify sixteen different meanings for *meaning* and try to resolve many problems with this term. Their book shows that they are using the positivist approach to word meanings attacked here. Consider this revealing quotation: "In most matters, the possible treachery of words can only be controlled through definitions" (206). What is it that creates this treachery of words? It is the fact that they can be used rhetorically, as explained in this volume.

10. Stanley Fish, *Surprised by Sin: The Reader in "Paradise Lost"* (Berkeley: University of California Press, 1967).

11. Stanley Fish, *Is There a Text in This Class? The Authority of Interpretive Communities* (Cambridge: Harvard University Press, 1980).

12. See especially Fish, *Doing What Comes Naturally*.

13. Michel Foucault, *The History of Sexuality, vol. 1, An Introduction*, trans. Robert Hurley (New York: Random House/Vintage, 1980).

14. Alasdair MacIntyre, *Whose Justice? Which Rationality?* (Notre Dame, Ind.: University of Notre Dame Press, 1988), 7. Fish's *interpretive community* can be seen as the agent for MacIntyre's *tradition* in that the interpretive community both embodies and extends a tradition. The conventions and constraints that define, and thus enable, the actions of the members become parts of the tradition within which they work.

15. In a helpful review essay comparing Fish's *Doing What Comes Naturally* to the work of MacIntyre, and entitled "The Unnatural Practices of Stanley Fish," (*South Atlantic Review* 55 [1990]: 87–97), Alan Jacobs put this point more succinctly: "A commitment to justice (and courage and honesty) is, in fact, internal

to a genuine practice, for without such a commitment, a practice will not strive for excellence, and thus will not 'contribute to human flourishing'" (91). I have hesitations about Jacobs's use here of the term *excellence*, but I like his formulation of an ideal that is not abstract.

2

Beginnings of the Creation/Evolution Controversy

In an early chapter of his 1955 book, *Apes, Angels, and Victorians: The Story of Darwin, Huxley, and Evolution*, William Irvine explains the origin of the theory of evolution by describing what Charles Darwin learned when he visited the Galapagos Islands during his world voyage from 1831 to 1836. I shall use the following paragraph from Irvine to introduce concisely the creation/evolution controversy and to illustrate a key rhetorical strategy used by evolutionists since Darwin:[1]

> The Galapagos Islands were the most *illuminating* lesson [of Darwin's world voyage]. His visit there seemed an actual journey into the biological past. . . . The landscape not only suggested evolution, *the facts demanded it.* Here [the] distribution [of animals and plants] reduced *the creation theory* to an *absurdity.* Each island had great numbers of species and varieties peculiar to itself, but related species and varieties, both in the archipelago and the adjoining mainland, differed from each other according to the magnitude of natural barriers between them. One could assume *"a creative power"* with an inveterate sense of localism or an illogical desire for busy work, but how much more *illuminating* to assume *an evolutionary force* producing, with *geographical separation,* increasing difference in the offspring of a common ancestor? While still in the islands he had actually described his data as "undermining the stability of species." From that time on the subject "haunted" him.[2]

According to this paragraph, Charles Darwin's experience in the Galapagos caused him to choose between two theories of the origin of species: creationism, which holds that God independently created each species of living things as described in the Bible, and the theory of evolution, which holds that different species have evolved naturally from a common ancestor. Darwin concluded from what he saw in these islands that the creation theory could not be true, and thus opted for the theory of evolution. To complete an outline of the creation/evolution controversy, it would only be necessary to trace the implications of contradicting the Genesis account of creation, to determine what role in nature (if any) is left for God, and to decide what policy to follow in teaching about the origin of species in U.S. public schools.

Irvine reveals the key rhetorical strategy used in this controversy by claiming that Darwin was confronted in the Galapagos with "facts" that "demanded" that he reject creation in favor of evolution. He writes that Darwin's direct encounter with these facts reduced the creation theory to "an absurdity," defensible only by positing a "creative power" (God) that is local, illogical, and wasteful. Instead of accepting this absurdity, Irvine twice indicates that Darwin was "illuminated" by the more sensible notion of "an evolutionary force," which produced these differences according to a universal, economical, and rational principle (which he abbreviates as "geographical separation"). In this paragraph, Irvine's principal strategy is to link the dichotomy of creation/evolution with a whole range of other evaluative dichotomies (unreasonable/reasonable, local/universal, wasteful/economical, dark/light, and so forth); within this linguistic net, he consistently links creation to the negative term and evolution to the positive term. He justifies this linguistic correlation by claiming that evolution matches "the facts" encountered by Darwin.

But what is the status of these facts? Just as poststructuralists have recently argued for new accounts of the relationship between words and meanings, philosophers and historians of science have recently developed new accounts of facts that problematize Irvine's claim. Such accounts contend that facts are not independent objects obvious to all observers; rather, facts take shape and assume significance only within a conceptual framework learned in human communities. These accounts suggest that scientific knowledge is socially constructed; it advances through the persuasion of a community of scientists rather than through the discovery of a simple match between natural facts and scientific theories. Such an account does not deny the existence of facts; it merely redescribes them as something other than obvious objects that are available to direct discovery.

By reenacting Darwin's discovery as a simple encounter with facts and an inevitable conclusion based on these facts, Irvine's paragraph implies that everyone else ought to agree: after all, "the facts" "demand" it. However, in the 140 years since Darwin first explained these facts in his 1859 book *On the Origin of Species*, many people have questioned his facts, rejected his conclusion, or both. Indeed, some people have attempted to develop an alternate account of these facts as direct proof that God created the species, thus arguing that the evolutionists themselves refuse to face the facts. How can one account for this fundamental disagreement? Irvine's paragraph implies that it turns on facts; those who do not accept the theory of evolution either do not know the facts or will not face them. However, creationists would not concede these points but would instead argue about what counts as a relevant fact. From the beginning, then, this fundamental disagreement typically involves a terminology battle between competing groups rather than a contest between people who face facts and others who refuse to do so.

Darwin and his followers have unquestionably persuaded millions of people around the world that the theory of evolution does indeed match the facts about the origin of species. Nor has this phenomenally successful persuasion gone unnoticed: countless studies have investigated the importance of Darwin's work, the specifics of his theory of evolution, and the effects of his rhetorical strategies. Indeed, in the history of biology and the philosophy of science, Darwin has assumed the status almost of a William Shakespeare and a full "Darwin industry" undertakes a wide variety of projects, such as printing all his writings, interpreting all his noteworthy passages, compiling concordances for his major works, collecting his voluminous letters, determining how others influenced him and how he influenced others, and evaluating his relationship to Victorian religion.[3] Rather than reduplicating any of this work, this chapter analyzes the key dichotomies used by Darwin and his antagonists in order to see how the disagreement between them was finally encoded in disputes about particular words. Instead of relying on Irvine's key term, *fact*, Darwin himself defended evolution primarily by linking it to the key term *science*, which his work redefined so as to dismiss creationism as unscientific, and therefore untrue. In effect, Darwin inherited a particular linguistic net of terms surrounding *science* and then rewove this net so as to include *evolution* and exclude *creation*. The chapter concludes by analyzing a famous creation/evolution debate: an 1860 encounter between T. H. Huxley, nicknamed "Darwin's bulldog," and Samuel Wilberforce, the bishop of Oxford. This classic debate weaves into the net a final dichotomy that is crucial to this controversy: it is a paradigmatic conflict between philosophy and rhetoric.

CREATION/EVOLUTION

The first dichotomy created by this controversy is the dichotomy between the two terms *creation* and *evolution*. Are they opposites? If so, can
they be reconciled or must one choose between them? The controversy itself exists only if their opposition is admitted and attempts are made to deal
with this opposition, either by reconciling the terms (and thus erasing the
opposition between them) or choosing between them (and thus allowing
the dominant term to erase its opposite). Perhaps the easiest reconciliation
is to assert that God used evolution as his method of creation, thus making
one term the end and the other the means. For those who resolve the difference so simply, the controversy is indeed hard to comprehend. However,
this resolution has been rejected by many creationists and evolutionists for
reasons that will become clear later on in this volume. The very fact that
one must decide what to do with these terms suggests that they are rhetorical tools rather than permanent realities: the terms themselves are mutually
defined by an unstable and hierarchical assertion of difference between
them.

One important difference is that these terms have often been used for different purposes by different human communities. Whereas for millennia,
priests and theologians have been using *creation* to talk about God's action
on the universe, scientists have been using *evolution* to describe the development of species from a common ancestor only within the last two centuries. Many people have decided how to reconcile the two terms or which
term to prefer by deciding whether to accept the definitions of scientists or
of priests.

In his article, "Introduction: Evolution and Creation," Ernan McMullin
traces the varying meanings of *creation* for the orthodox Christian tradition,
including its uses by such important figures as Saint Augustine, Saint Thomas Aquinas, René Descartes, and Sir Isaac Newton.[4] Although Christian
thinkers have often differed on the meanings of the term and the details of
creation, they all conceive of God as their creator and themselves as his
creations (or his creatures); this relationship motivates them to worship
God and motivates his actions toward them (or, in many accounts, his in-
carnation among them as Jesus Christ), necessary actions to allow them to
know him at all. The term *creation* appears in the first verse of Genesis: "In
the beginning, God created the heavens and the earth." A synonym appears
as the first sentence of the most common Christian creed: "I believe in God
the Father Almighty, maker of heaven and earth." In his book about the
Christian doctrine of creation, *Maker of Heaven and Earth*, Langdon Gilkey
writes: "The idea that God is the Creator of all things is the indispensable

foundation on which the other beliefs of the Christian faith are based."[5] It is hard to overestimate the importance of this term to Christianity or the loyalties it evokes in Christian theologians and ministers.

These strong loyalties are shared by many lay Christians. Every reader of the Bible has encountered the term *creation* within its first verse. The Book of Genesis goes on to explain that God worked for a series of "days" and that he created living things "after their own kind, bearing seed in themselves" so they could reproduce. Even before the doctrine of biblical inerrancy was developed in print, many believers in the Bible felt bound by this account of the origin of species and called it by the name *creation*.[6] Thus, in many contexts the word carries religious valences for its users.

The term *evolution* has a significantly different history and a different set of valences. McMullin argues that several ancient ideas of Hippolytus, Democritus, Empedocles, and Aristotle about how "the universe might have gradually come to be as it is" roughly correspond to our contemporary notion of evolution.[7] The term's actual use in English has been traced by Peter J. Bowler from the Latin *evolutio*, meaning "the act of unrolling." Bowler explains that the word was first used in English to describe the development of an embryo into an adult, but its meaning gradually shifted from this sense of unfolding according to a preset plan into its modern sense of changing without any particular goal.[8] *Evolution* is not a word that people encountered regularly in church, family, or school—or in the first sentence of the Western world's most powerful book.

Bowler reports that the current usage was not introduced or even popularized by Darwin. The word *evolution* was first used in its modern sense by the geologist Charles Lyell in 1832. Darwin himself did not use it in the *Origin* to describe his theory (the book ends with the word *evolved* but does not otherwise use the term), and it is hardly mentioned in the extensive debates about the *Origin* among Darwin's contemporaries.[9] Similarly, the phrase *theory of evolution* was first used, not by Darwin, but by Herbert Spencer in 1852; Darwin did not accept this term for his theory until much later.[10] Darwin called his account the *theory of descent by modification* and thought of his key term as *natural selection*, but Spencer (who was a student of rhetoric) felt that *theory of evolution* was more effective; he also coined the vivid phrase *survival of the fittest* to intensify the more neutral *natural selection*.[11] Largely due to Spencer's influence, the term became widespread about 1870; T. H. Huxley wrote the first encyclopedia article about it in 1878 for the *Britannica*.[12] An irony of the creation/evolution controversy is that even Darwin himself did not feel much attachment to the main term that was later linked to his name.

How did *evolution* develop its importance? Not by appealing to Christian believers from all ranks and backgrounds over 2,000 years, but by seeming important to a small group of scientists in Western Europe who lived in the mid-1800s and who came to see it as a scientific explanation of natural phenomena. They became committed to the term, developed its meanings, and disseminated it to the general populace. The surprising thing is that a word that was introduced and debated by a small number of scientists around 1850 has been able, in certain uses, to replace a word that lay at the heart of a major world religion for 2,000 years. In response to a request to explain the origin of species, a vast number of people now say that "different species evolved" rather than that "different species were created by God."

One method by which some evolutionists have established their theory is to contrast a scientific theory to a religious belief, a consciously known truth to an unconsciously accepted dogma. Although there are many important differences between science and religion, professionals in both areas of knowledge claim to provide true statements, which somehow correspond to reality. As long as these truth claims can be applied to different spheres, science and religion do not obviously conflict. However, in the case of the origin of species, the truth claims of science and religion have been taken by many people to be directly conflicting. Those who have chosen the theory of evolution over the creation account have often felt they were also making a more important choice to believe either the scientists or the theologians and priests.

GOD/NATURE

Since the Enlightenment, the choice between science and religion has been further complicated by another important dichotomy: the difference between God and nature. In brief, the Enlightenment originators of the scientific method (almost all of whom were devout Christians) redefined and reversed the God/nature dichotomy as conceived and revised through history by orthodox Christianity in such a way that nature became the primary term instead of God: nature became a proof of God rather than God being the origin of nature. This shift in the God/nature dichotomy set up a relatively recent series of linkages between religion and science, belief and knowledge, and ultimately, creationism and evolution that contributed to the widespread acceptance of evolution.

In his book *At the Origins of Modern Atheism*, Michael J. Buckley focuses on the following problem: whereas people once considered the term *atheism* so derogatory that they would not allow themselves to be called atheists, this term gradually became acceptable enough for many people to use it even to describe themselves. Buckley's book studies how this term was reconceived as part of a shift in worldview that made atheism an acceptable and common intellectual position. The book is, therefore, an analysis of terminology battles about the meaning and use of *atheism*.[13]

Buckley explains the shift in this term by showing how Enlightenment thinkers such as Descartes and Newton gradually shifted the study of various ideas associated with God from theology to philosophy, thus paving the way for God's role in the universe to be diminished in exact proportion to the extent that scientific explanations of the universe increased in precision and predictive power. These thinkers conceived of science as an effort to explain mechanically many of the phenomena that people used to assume were inexplicable and thus had attributed to God.

Buckley points out that the Enlightenment thinkers accomplished this shift entirely against their intentions. Both Descartes and Newton explicitly discussed the relationship of their ideas to God and showed why God was still necessary in spite of the success of their efforts. Descartes made God a logical necessity behind his universal mathematics; Newton made God the author of his universal mechanics.[14] However, both arguments mentioned God only after their scientific accounts were complete, not as an integral part of these accounts. With time, it came to be seen that neither mathematics nor the physical sciences needed God to explain any of their operations. As a result, such attributes as eternity and infinity that used to belong solely to God were transferred to nature and to matter itself.[15]

Buckley thus traces a reconception of the difference between God and nature.[16] Both Augustine and Aquinas made it clear that the universe was a contingent affair; its very existence depended on the prior existence of God. For example, McMullin quotes Augustine that "the universe will pass away in the twinkling of an eye if God withdraws his ruling hand."[17] However, Newton and Descartes managed to account for many important phenomena in strictly mechanical and mathematical terms, thus lessening the sense of God's ruling hand and instead including God almost as an afterthought. With time, this sense of God's supplementarity became obvious and the dichotomy shifted so that nature came to seem more fundamental and real than God. This new method of explanation became known as *science*, and it took as its fundamental task the construction of explanations according to natural rather than supernatural or human causes and effects, thus distinguishing nature from both humanity and God. To account for something scientifically came to mean to explain it in terms of natural processes rather than in terms of human choices or God's intervention.

Based on the astounding and admirable success of science in developing useful accounts of nature over the last 300 years, some people have concluded that the scientific method is the only sensible way to deal with all kinds of human needs and problems, and thus that there is no need for any discussion of God. Buckley does not share this conclusion, however. He contends that theology made a mistake around 1700 by ceding its territory in this manner to philosophy: the very desire to be rational rather than fideistic caused Christian theology to forget that its major focus is not impersonal and abstract explications of God as an idea, but individual and collective experiences of God as a person, as the historical figure of Jesus Christ.[18] Buckley's thesis can be taken as a call for theology to return from a study of abstract ideas to a study of the people behind those ideas.

Buckley's book also suggests the importance of the next major work for the creation/evolution controversy: William Paley's *Natural Theology* (1802). This highly influential book took the next step after the Enlightenment rationalists; it attempted to explain how a Christian scientist should view scientific discoveries in relation to religious commitments, how to deal with the difference between nature and God. The book argues that the discoveries of science reveal the methods used in nature by God and implies that the more one learns about these methods, the greater will become one's admiration for God. In a now-classic analogy, Paley compares nature to a watch and God to a watchmaker.[19] The analogy works as follows: If someone were to discover a watch lying on the ground, he or she might decide to open it up and discover the workings of all the internal parts. The more the person found out about these workings, the more he or she would know scientifically about the watch. But how would the person account for the existence of the watch itself? He or she could only conclude that it was designed by an intelligent designer, a watchmaker, and the more the person studied the watch itself, the more he or she would come to admire its designer.

Through this analogy Paley reconceived God and nature as two separate topics of study: the how of nature and its why, its workings and its origin, its laws and its lawgiver. By distinguishing these two topics, Paley was able to convince a generation of scientists that their work was adding to the glory of God even as they were diminishing his actual involvement.[20] The title of his book itself clarifies the point: *natural theology* is the effort to account for God through nature. John Durant defines it as "the study of the existence and attributes of God as manifested in the works of nature."[21] Its opposite is *revealed theology*, the belief that one learns about God only through direct revelation. To think that God can be studied in nature is already to shift the terms of the God/nature dichotomy; in Paley's title, the primary term is already being usurped by what Jacques Derrida might call a "dangerous supplement."

What Darwin did was to complete this usurpation by extending the domain of science from the physical sciences to biology. Even after much of the physical world had been explained in terms of natural causes and effects, many people remained convinced that the world of living things was too complicated to be explained mechanically. After all, they reasoned, the laws that govern life must be infinitely complex; even if mathematics could explain the motion of a planet, it could not account for a bird or a flower. Of course, science thrives on just such challenges. Certainly Darwin decided to accept it in dealing with the problem of the origin of species, which he considered the hardest problem of all biology and described as "the mystery of mysteries."[22] In effect, Darwin performed a test of Paley's natural theology by trying to explain how species are formed and to see what would happen to our understanding of why.

What Darwin discovered was that in this case, the "how" *was* the "why." In his book *Charles Darwin and the Problem of Creation*, Neil C. Gillespie argues that Darwin gradually came to see that his account had to eliminate completely the notion of creation in favor of a mechanism that would work as a natural cause. Based on the shifting senses of *God* and *nature*, Darwin realized that any argument in which God had something to do with the origin of species would not be a natural explanation. This conception of the dichotomy required that God's role must not involve natural causes and effects or it would become impossible to distinguish the works of God from the works of nature; if God was not nature, then everything that was nature could not be God. In Gillespie's terms, Darwin concluded that a scientific account must change its notion of law "from law as divine will to law as no more than observed regularity of behavior"; whereas "the old science invoked divine will as an explanation of the unknown, the new postulated yet-to-be discovered laws."[23] Darwin set out to discover those laws, which were later regularized as the theory of evolution.

To see Darwin himself positing these relations between key terms, consider this sentence from the *Origin*: "It is so easy to hide our *ignorance* under such expressions as the 'plan of *creation*,' 'unity of design,' etc., and to think that we give an *explanation* when we only *restate a fact*."[24] This quotation explicitly links the term *creation* to the devalued terms *ignorance* and *restatement*, implying that Darwin's new account is an example of their opposites, the valued terms *knowledge* and *explanation*. By explaining the origin of species in terms of natural selection as opposed to special creation, throughout the rest of his book Darwin explicitly eliminated *God* as a term, thus showing that the process of speciation has its own internal motor or cause. He argued that species develop when the animals that are the most successful in adapting to their environments pass on their traits to the greatest number of offspring, a process that eventually leads to the formation of new, better-adapted species and the extinction of old, worse-adapted species. In Darwin's account, God became an unnecessary term, or to put the case more strongly, a term that would ruin the explanation. Gillespie concludes that Darwin had to eliminate the notion of creation from his account of the origin of species in order to make biology scientific.[25]

In this case, Darwin thus contradicted Paley's design argument. Rather than the laws of nature proving the existence of a designer, in this case the laws either usurped the designer or failed as laws. To return to Paley's watch image, Darwin was able to explain in natural terms, without any reference to God, not only the most complex and challenging watch in the world, but the particular watchmaker: natural selection.

One of the principal efforts of British and American scientists (most of whom were Christians and many of whom were clergymen) for the rest of the nineteenth century was to figure out what role was left to God if Darwin was correct, a problem they called "the God of the Gaps." Their efforts have been summarized in two books about many of the scientist-theologians who tried to accept Darwin's insight yet retain some role for God. Because of the importance of Paley, the question continued to focus primarily on the problem of design.[26] God kept receding from these thinkers, but they kept trying to keep God from disappearing completely.

Even Darwin himself apparently faced this problem. Besides tracing Darwin's notions of creationism and science, Gillespie also tries to make sense of his abundant use of theological language, including the term Creator, in a book whose purpose was to dislodge the belief in special creation; he concludes that Darwin himself believed that life was first created by God.[27] Apparently, the rejection of Paley's design argument caught even Darwin in another troublesome dichotomy: that of chance versus design, the idea that to eliminate God as the creator of the universe was to assert that life had developed strictly by happenstance. Darwin wrote a moving letter to the American botanist Asa Gray in which he frankly confessed that he could not be comfortable with either idea and was deeply troubled by this problem, which he felt his own theory had created.[28] He apparently felt that he wanted to retain God, but did not know quite how.

What was a creationist to do in the face of Darwin's account of the origin of species? In effect, Darwin's theory of evolution redefined science and extended its domain so as to explain biology through natural causes and effects by methodologically excluding God and the notion of special creation. During this same period, geologists were applying Charles Lyell's principles of uniformitarianism and concluding that for many reasons, the Genesis account could not be true.[29] Many people were also losing faith in the Bible as a result of contemporary social problems and upheavals in industrializing Britain.[30] Believers in God saw all these forces combining to undercut the authority of Genesis and concluded that God was under attack, in part by the theory of evolution.

How was one to defend God? There were doubtless many ways, but a popular one was to reject the dichotomies posited by Darwin and the links he had established between them. If believers did not agree that creationism was absurd and that God had disappeared from biology yet also did not want to reject science, they could try another tactic of defense, which has been used by many creationists since Darwin. They could try to redefine *science* and reverse its historical shift in meaning by linking it to other important dichotomies—such as knowledge/belief and proof/assumption—in an attempt to question whether evolution really qualified as science, and thus to defend the validity of creationism. Thus did the implicit linguistic contests by association end and the explicit terminology battles begin.

SCIENCE

Thus far this chapter has traced some of the crucial terms linked to *science* since it began to develop its current meanings during the Enlightenment (or, as it is also aptly called, the Scientific Revolution). Among these terms are *nature, explanation, fact,* and *law.* However, long before 1700—in fact, at the dawn of canonical Western civilization, with the Greeks—early synonyms for the English term *science* were being situated in other linguistic nets and constructed in other violent hierarchies. Larry Laudan suggests that perhaps the earliest of these hierarchies was the Greek dichotomy of episteme/doxa, the difference between knowledge and mere opinion. Laudan explains that in the *Posterior Analytics,* Aristotle specified the conditions under which knowledge could be rendered so different from opinion that it could qualify as science, the highest form of knowledge. The most important of these knowledge qualifications was certainty.[31] To use Aristotle's distinction (which he borrowed from Plato), any statement held by anyone can count as an opinion or a belief (whose Greek work is related to the negative English term *dogma*), but only certain statements can qualify as knowledge. On what basis is this certainty to be determined? The most famous basis—the one shaped centuries later into "the scientific method"—has been to eliminate all subjectivity by submitting the statement to a rigorous process of doubt and testing so that it can be upheld with objective proof. In addition to the terms that already have been traced, the net surrounding *science* thus came, by the time of Darwin, to include the terms *knowledge, objectivity, certainty,* and *proof.*

Some of the current valences of *science* can be suggested by its use as an adjective in such phrases as *scientific fact* and *scientific theory.* What is implied by a *scientific fact* as opposed to a *fact?* The addition of the adjective suggests that the latter is superior to the former because it has been arrived at more rigorously than other facts. It has been proven objectively by a community of scientists, who search for the truth about nature. This community calls their endeavor *science* and regularly discusses their work in terms that implicitly devalue other kinds of work, such as work that does not begin with doubt, test its formulations rigorously (at least not in the same way), nor characterize its primary object as a search for truth.

These valences for the term *science* are maintained in part by the scientists themselves, who carefully limit its use. They make efforts to assure that only those who meet certain requirements may legitimately apply it to their work, which must receive the approval of their particular (and often small) scientific community. It is well known that in the past two centuries, many academic disciplines have tried to define their tasks so that they can meet these standards and call their work by this name (other disciplines have undertaken similar protective moves for their own key terms). The fact that some disciplines attempt to police and others to appropriate the term *science* suggest that it has considerable linguistic power in Western culture. It retains part of this power by associating itself with positive terms and implicitly devaluing their opposites.

From the time of the Scientific Revolution, when they first called their work "natural philosophy" or "new philosophy," until today, when the term *science* has gradually developed its present meanings, those communities that label their work with this term have often defended their goals and methods by appealing to their considerable technological achievements as proof of the value of this method and by using a rhetoric that equates their findings with discoveries of the truth. Thus, grade school students learn about various "discoveries": Nicolaus Copernicus discovered that the earth revolves around the sun, William Harvey discovered the circulation of the blood, Albert Einstein discovered that matter is another form of energy, and so forth; they also learn that all these discoveries improved human life. These stories are usually told as the replacement of old, incomprehensible, and possibly even dangerous ideas by new, intelligent, helpful ideas that exactly match the world in which they live.

The quotation by Irvine at the start of this chapter uses such a rhetoric. It is also the rhetoric in which many students first learn about Darwin. However, an unusual thing sometimes happens when the story of Darwin and evolution is told to schoolchildren in these familiar terms: some students refuse to believe this story about Darwin yet have no trouble at all believing similar stories about Copernicus and Harvey. How would a scientist account for this? Some would probably use the same terms that were quoted from Darwin: as an act of "ignorance," a use of "expressions" such as "special creation" in place of "explanations" of the "facts." Some scientists blame it on the power of particular religious beliefs, which they think act as obstacles in the way of learning. If it were not for these false beliefs, they imply, scientific knowledge would achieve greater acceptance and the world would be a better place.

Some people do reject evolution because it conflicts with their ideas about God or the literal truth of the Bible. However, their rejection of the theory of evolution does not prove that they are enemies of science as they conceive it. In fact, they often claim they are defending science *against* evolution. To support their position they use a strategy that has been very common since 1859: they reject evolution on the grounds that it does not qualify as *science*. They use this argument because they admire science; they want to retain its positive valences and its power as a term. As a result, they continue to assert that science discovers the truth about nature by eliminating Darwin from their list of notable scientists.

In recent years a number of philosophers and historians of science have also been focusing on the term *science* and its power, but in quite a different way: they have attacked the conception that science discovers the truth about nature as revealed in preexisting facts. They have argued instead that science progresses as a result of a mutually constitutive relationship between fact and theory. In this view, both a scientific theory and the facts confirming that theory emerge together as functions of each other rather than as separate entities with independent existence which need only to be correlated. This new account considerably changes the linguistic net that links *science* to *knowledge*, *certainty*, and *proof*.

The most influential of these new accounts, Thomas Kuhn's *The Structure of Scientific Revolutions*, gives an example in Galileo Galilei's "discovery" of the pendulum.[32] When Aristotle studied motion, he concluded that an object that swings back and forth on a string until it comes to rest can be classed as a body that is falling with difficulty. However, when Galileo studied the same phenomenon, he conceived of this object as a pendulum. Kuhn asks whether it makes sense to say that the object was a pendulum all along and that Aristotle failed to see it correctly. Kuhn decides (with considerable uneasiness) that such a view of Aristotle and Galileo makes no sense. Even though history books report that Galileo discovered the pendulum as if the pendulum had always existed and Galileo was the first person to notice it, the discovery can only be accounted for as a new way of seeing falling bodies, which apparently was unavailable to Aristotle. Kuhn concludes that a scientific fact is always the product of a certain way of seeing and agreeing about a natural phenomenon. The group that must agree on this fact is a community of scientists. Because later physicists preferred Galileo's conception of a pendulum to Aristotle's conception of an uneven fall for a variety of complex reasons, including the former's greater usefulness as an account, this conception attained the status of a fact and thus became associated with certainty, knowledge, and truth. Aristotle's conception was subsequently reported as an error, an opinion, or an incorrect view, to be replaced by Galileo's correct one.

In the years since Kuhn's seminal work, many similar studies have been undertaken, including Paul Feyerabend's account of how Galileo persuaded the Catholic Church that the earth moves around the sun and Bruno Latour's explanation of how Louis Pasteur persuaded the French that invisible beings called microbes cause diseases and that expensive and inconvenient public health measures are sometimes necessary in order to kill these microbes before they kill more people.[33] These studies have all agreed that the linguistic net surrounding science needs to be revised; indeed, the knowledge that counts as science is not provable or true in any simple sense. They contend instead that the decision to call something *scientific* is made by various communities, beginning with a community of scientists; this decision does not depend on matching "the word" to "the thing" nor the fact to the explanation, but on persuading people to accept key definitions and conceptions, to decide what counts as a fact and what as an explanation. Such authors admire science and want to explain it as a human achievement rather than a mechanical discovery of abstract truths.

In the case of Darwin, this rhetorical aspect of the word *science* is crucial. A terminology battle about the meaning of *science* was not invented later by creationists to fight against evolution; the problem was faced squarely (and even conceded) by Darwin himself, who admitted that he was unsure whether his work qualified as science. The problem was hotly debated after the *Origin* came out. In fact, evolution by natural selection was not generally accepted as a scientific explanation until twentieth-century discoveries in genetics and cell biology provided a more solid mechanism (DNA) through which the workings of natural selection could be understood than Darwin had provided. Darwin's efforts to persuade his contemporaries that his famous work qualified as science suggest that the term *science* is not the name of an absolute reality but rather a tool used for purposes of persuasion that changes over time.

DARWIN'S *ORIGIN OF SPECIES* AS SCIENCE

During the years when Darwin was receiving his scientific training at Cambridge,[34] the governing notion of science was taken from Francis Bacon, who held that science is inductive rather than deductive, that it gathers facts and then generalizes them into a scientific theory.[35] To oversimplify, the view was that science ought to limit itself to statements that could be proven through close observations; any statement that could not be thus proven could be dismissed as a mere opinion or belief.

Darwin received a vivid lesson in how science works one day while on a field trip to Wales with one of his mentors, the Cambridge clergyman/geologist Adam Sedgwick. Darwin excitedly told Sedgwick about an amazing geological discovery he had recently made: a tropical shell lying in a gravel pit near his home. Sedgwick dismissed this great scientific find by saying that someone must have thrown it there and told the disappointed Darwin that "if really embedded there it would be the greatest misfortune to geology, as it would overthrow all that we know about the superficial deposits of the midland counties."[36] Through the very effective pedagogical method known as embarrassment, Darwin learned that science means something more than reaching a conclusion based on one fact. Perhaps he learned that science must be able to recognize a significant fact in the first place. It must also account for a whole group of facts and organize those facts into a reasonable and (at least in principle) provable statement.

When Darwin applied his scientific training to the problem of the origin of species, however, he scarcely could have followed Bacon's formula. First, he had to decide what counted as a fact. While in the Galapagos, for example, he was not collecting facts for a great future work, but only noticing and collecting species of plants and animals. While traveling to the different islands, he began to notice similarities and differences between the species he found there. He was able to pick out these similarities and differences because he had been trained to observe them.

Among his observations, perhaps the most famous concerned the differences between the bills of finches on neighboring islands. (Only a careful naturalist making a record of such details as the size and shape of finch bills would probably have noticed these facts and attempted to decide what they meant). With time, Darwin began to interpret the distribution of species as part of a significant pattern that related to the problem he had begun to frame about the origin of species. This problem was also a product of his particular task; he was trying to keep track of all the species he found, and he had to keep adding new entries to his list as he went from island to island. He began to wonder why each of these islands would have slightly different species and why God would have possibly busied himself with making such intricate and minor changes on so many nearby places. Throughout the entire process, his observations were constituted as significant facts within the problem he had framed. They became facts worthy of notice in relation to the pattern he began to see because of the task he was trying to do as a trained naturalist.

After returning from his voyage on the *Beagle*, Darwin amassed a huge group of facts over years of study (facts that were similarly constituted by his interpretive labors), but these facts proved very difficult to organize into a verifiable generalization. Indeed, his thinking about this problem stretched over twenty-seven years: from not long before his *Beagle* voyage in 1831; through 1842, when he wrote a 35-page sketch of his ideas; 1844, when he expanded this to a 230-page draft; and then to 1858, when he decided to finish a one-volume draft and publish it.[37] He later reported that he hit on his theory, not by marshaling naked facts, but through reasoning by analogy from artificial selection to the processes of nature (thus the origin of his term *natural selection*). In brief, he reasoned that just as humans can consciously control the breeding of animals and plants to improve their livestock and crops, nature must use some method to determine which varieties flourish and which ones shrink. He found this method in the sixth edition of Thomas Malthus's famous Essay on the Principle of Population (1828), which described the competition between an exponentially increasing population and a limited supply of vital resources. Darwin did not "discover" natural selection by objectively encountering facts that no one could miss, but rather by completing conceptual labors such as observation, correlation, and interpretation—labors of many different kinds.

The same point can be made about *proof*. When Darwin attempted to provide the Baconian proof for this theory of natural selection (proof that, incidentally, would still emerge only through the same processes of selection and interpretation—not from a direct match between statement and reality), he could not show, either in the fossil record or in the laboratory, that one species had ever changed into another. Thus, Darwin decided to use a different method to prove his theory. This method is summarized by Robert M. Young: "Darwin's task [in the *Origin of Species*] was to explain *away* the *lack* of evidence while repeatedly stressing the greater plausibility of his theory over that of special creation."[38] Rather than providing indisputable proof for evolution, Darwin decided to construct a plausible narrative of his thought processes. His compelling narrative has provided sufficient proof of this theory to persuade millions of readers to accept it. Another irony of this controversy (or, more accurately, an unavoidable product of this redescription of it) is that Darwin deliberately appealed to rhetoric/persuasion in an effort to make his theory scientific/philosophical/true.

After the publication of the *Origin*, in November 1859, Darwin anxiously awaited the reaction of the scientific community. He explained his anxiety in several letters to close friends. To Asa Gray he wrote: "I am quite conscious that my speculations run quite beyond the bounds of true science."[39] To another friend he admitted that "the change of species cannot be directly proved [and] . . . the doctrine must sink or swim according as it groups and explains phenomena."[40] When Henry Fawcett, a Cambridge Professor of Political Economy, published a favorable review of the *Origin* in *Macmillan's Magazine*, Darwin wrote him a grateful letter, to which Fawcett responded in part:

> [I want] to point out that the method of investigation pursued [in your book] was in every respect philosophically correct. I was spending an evening with my friend Mr. John Stuart Mill, and I am sure you will be pleased to hear from such an authority that he considers that your reasoning throughout is in the most exact accordance with the strict principles of logic. He also says the method of investigation you have followed is the only proper one for the subject.[41]

Darwin was greatly relieved to hear of this reaction by England's premier expert on logic. He replied to Fawcett that he was grateful to hear about Mill's reaction: "Until your review appeared I began to think that perhaps I did not understand how to reason scientifically."[42] These letters show Darwin conducting his own internal inquiry into the correct use of the word *science*. He wanted to call his work *science* but wondered whether it qualified. Thus he alternated back and forth between definitions of the term and asked his friends what they thought.

With time Darwin and his associates became convinced that his work was scientifically sound. However, this was hardly the reaction of the rest of the scientific community or the world at large. After a study of many contemporary reactions to Darwin, David Hull has concluded that very few people initially accepted Darwin's method or his conclusions, including the three foremost British philosophers of science: William Whewell, John Herschel, and even J. S. Mill, who ultimately rejected the theory of evolution although he admitted that it was scientific.[43] Henry Fawcett provides some insights into this rejection in a letter to Darwin: "It is easy for an antagonistic reviewer, when he finds it difficult to answer your arguments, to attempt to dispose of the whole matter by uttering some such commonplace as 'This is not a Baconian induction.'"[44] For the thesis developed here, this quotation works as a two-edged sword. From his own perspective, Fawcett feels that many experts refused to accept Darwin's method because they were too stubborn; intimidated by the truth of Darwin's arguments, they refused to concede the correct meaning of *science*, retreating instead to a commonplace about Baconian induction. From the perspective of Darwin's opponents, however, restating this reservation was hardly refusing to face the truth. Darwin was violating the term *science*, as understood and used by their communities; they felt obligated to protect the term by rejecting his use of it. Darwin and his opponents wanted to use the valorized term *science* differently, by differently weaving its linguistic net.

In the light of this opposition, how did Darwin's theory ultimately become accepted as science? Several studies have attempted to account for its successful persuasion. In his essay, "Darwin's Metaphor: Does Nature Select?" Robert M. Young argues that Darwin's theory was aided by the very ambiguity of his central principle. Young indicates that the metaphor of natural selection was a subject of continuous controversy from 1860 to 1930; the question was how *nature* could *select*.[45] Darwin could not provide a mechanism to answer this question. Instead, he appealed to the principle of uniformity itself, saying that no other explanation would work, given the uniform laws of nature. Young reads the book as Darwin's effort to elicit "faith in the philosophical principle of the uniformity of nature," and adds: "Whenever he was really in trouble, he adopted the same tactic as [Sir Charles] Lyell, [Robert] Chambers, and [Baden] Powell had done—he appealed to the very principle which was at issue."[46]

The circularity of this argument was not lost on some reviewers. For example, Samuel Wilberforce asked in one review why nature should be so uniform in its workings except when it changes so as to develop new species.[47] However, this very circularity allowed the theory to remain scientifically possible until a more detailed explanation could be found. Young concludes that the metaphor worked because it was a "frail reed," able to bend with the winds of naturalism and, ultimately, to "bring the earth, life, and man into the domain of natural laws."[48] According to this interpretation, Darwin had lost the battle but won the war, at least until it could be taken over by others. In another tale that could be told as a complex case of persuasion, geneticists and cell biologists in the early twentieth century were eventually able to provide mechanisms (such as the double helix found by James Watson and Francis Crick) that explained natural selection and helped shift the meaning of the term *science* so that the theory of evolution ultimately fulfilled the criteria of science for the majority of the scientific community. In the years since the discovery of these mechanisms, evolution has become almost universally accepted among scientists as a classic predictive theory and has led to many other biological discoveries and innovations.

Other factors besides the ambiguity of Darwin's metaphor help to explain how evolution was able to link itself to the powerful word *science*. In "Evolutionary Biology and Ideology: Then and Now," Young explains some overtly political reasons why evolution succeeded as science.[49] He argues that Darwin defended natural selection throughout his life because he accepted its implications of inevitable competition, whereas scientists who felt differently about competition (and capitalism, its political/economic corollary) rejected natural selection as the mechanism of evolution. For example, A. R. Wallace argued that natural selection was used by many thinkers to blame "nature for man's inhumanity to man and [to take] a fatalistic view about the impossibility of radically restructuring society" (184). In direct contrast to Wallace, Herbert Spencer supported natural selection even more avidly than Darwin because he liked its implications for a free market economy (184). T. H. Huxley decided to limit natural selection to the earlier stages of evolution, arguing that when evolution reached the stage of humans, it should stop; ethics should then take over in governing human interactions (185). These political implications were visible to many others besides these scientists. For example, Karl Marx and Friedrich Engels saw natural selection as an economic ideology: Engels wrote that evolution was Thomas Hobbes's "war of all against all" applied to nature (203 n.N), and Marx so admired the *Origin* that he wanted to dedicate *Capital* to Darwin (although Darwin politely refused and evinced surprise that Marx should see any parallels between their ideas) (185). Many other stories could be told about the politics of evolution as science. According to this interpretation, its scientific status is, in part, quite literally political.[50] To call it political is not to attack the theory, but to contend that it has attained its stature in the same way as all scientific theories: by meeting important human needs as a function of particular values and beliefs.

In a recent book Peter Bowler explains the triumph of evolution in another way: as a story of explicit strategies of persuasion.[51] Bowler agrees that the theory of evolution has created a scientific revolution in biology, but he believes that any account describing Darwin's work as a heroic effort to advance the truth is a myth. He writes: "All cultures have myths about their origins, designed to legitimate their members' assumption of a privileged status. The scientific community is no exception, and Darwin has become the hero or founding father in the creation-myth of modern evolutionism."[52] Bowler accounts in part for the success of evolution by describing a creation-myth believed by scientists themselves. Scientists as a community think of their work as valuable because of the stories they hear and tell. Among other stories, the story of Darwin's discovery of evolution gives them reasons to continue their own scientific work and to advance science as a crucial value system in their own lives, just as the Genesis story functions as an important source of value and meaning in the lives of many Christians. The triumph of evolution for scientists thus becomes another story of the triumph of science itself.

Bowler also attributes much of the success of evolution to T. H. Huxley; he argues that Huxley did not agree with Darwin on most key concepts (including natural selection and the social implications of evolution) yet nevertheless supported evolution for reasons of his own, especially his desire to advance science education. Bowler calls Huxley "an exceptionally skillful politician of science" who helped Darwin as part of a strategy; he adds, "To succeed in the game of scientific politics, Darwin had to play his cards very carefully."[53]

John Angus Campbell has also written extensively about rhetorical aspects of Darwin's defense of evolution. In an article entitled "The Invisible Rhetorician," Campbell traces Darwin's efforts to use letters to recruit other scientists to speak for him and to preserve his image as a detached man of science by suggesting to Huxley and Gray how to conduct public battles for him while he remained at home.[54] Campbell would agree with Bowler's conclusion: "It was through persuasion and through success in the politics of science that Darwinism came to dominate British biology."[55]

These various interpretations of the reasons for Darwin's success suggest that he used many methods to construct the theory of evolution as science in opposition to the unscientific idea of creation. The triumph of evolution is not a simple matter of obvious facts nor of universal definitions of terms. Rather, evolution has been constructed as science, and the term *science* itself has changed over time through many concrete efforts at persuasion. As a result of this linguistic construction, *science* is now associated in many minds with *knowledge, certainty, proof, fact, law, explanation,* and other positive terms, which link it with the powerful Western term *truth*. In short, it has succeeded by convincing many people that it is true. To analyze some of the dimensions of this persuasion is not to deny the truth of evolution or to suggest that it should not be believed, but only to show that a conviction of its truth—and of the truth of any proposition—is, inevitably, the result of effective persuasion.

THE HUXLEY/WILBERFORCE DEBATE

In the famous debate between T. H. Huxley and Samuel Wilberforce, we can see how such convictions of truth are created through particular rhetorical strategies. As this incident and its various retellings show, this controversy regularly pits competing definitions of terms against each other and includes as one key opposition a conflict between rhetoric and positivist philosophy.

In June 1860, Professor Huxley and Bishop Wilberforce had a brief clash before 700 spectators at the Oxford meeting of the British Association. The story of this clash has attained the status of a cultural myth. It is often used in debates about academic freedom, where it reenacts Socrates's conflict with the Athenians and Galileo's battle with the Inquisition. It prefigures the Scopes "Monkey" Trial (the topic of the next chapter). It has been a topic for British Broadcasting Corporation specials about Darwin and the progress of science. It is also the introductory anecdote in many histories of evolution. This and other episodes in this controversy are regularly retold almost as modern epics, with an evolutionist cast as the hero and a creationist as the villain (or vice versa), who must be overcome if civilization itself is to be kept from destruction. Why is this so? Some tellers see the hero as philosophy and the villain as rhetoric, and they tell the tale repeatedly because they fear that rhetoric will lead to the collapse of civilization.

The present chapter began with William Irvine's account of what Darwin learned while in the Galapagos. Although this is an important moment in Irvine's book, it is not the most vivid. Irvine begins his book with a chapter entitled "Revolution in a Classroom," in which he tells the story of the Huxley/Wilberforce debate. He places this interesting episode at the front to get the reader hooked. Irvine's entire chapter deserves a close reading for the dichotomies he uses to contrast Wilberforce and Huxley, but the following analysis will focus only on some important passages.

Irvine's overall strategy is to link the bishop, "Soapy Sam," with the qualities historically associated at various times with rhetoric, and the scientist, who serves as "Darwin's bulldog," with qualities linked to philosophy. The first paragraph alone (which describes in five sentences the setting into which Huxley came for the debate), makes many implicit contrasts: the cleric Wilberforce works in Oxford, in rural surroundings reminiscent of the Middle Ages where "ideas [are] as ivy-covered as the building, and minds as empty and dreamy as the spires and quiet country air"; the scientist Huxley works in the busiest section of nineteenth-century London, where the streets are "as crowded and busy as Professor Huxley's own intellect."[56] Within a few sentences Irvine has already established the dichotomies of rural/urban, quiet/bustling, empty/full, thoughtless/thoughtful, inactive/active, medieval/modern, religion/science, and Wilberforce/Huxley, in each case aligning the implicitly devalued first terms against the valued second terms.

As the chapter continues, Irvine weaves his linguistic net to include other dichotomies that have been used at various times in the conflict between rhetoric and philosophy. The dichotomies and valences all come together when Irvine finally introduces Wilberforce, just before the bishop begins to speak to the assembly: "Bishop Wilberforce, widely known as "Soapy Sam," was one of those men whose *moral and intellectual fibers* have been permanently *loosened* by the early success and applause of a distinguished undergraduate career. He had thereafter taken to succeeding at easier and easier tasks, and was now, at fifty-four, *a bluff shallow good-humored opportunist* and *a formidable speaker* before an *undiscriminating crowd*" (6, italics added). In a series of equations that have been used successfully since the time of Plato to attack rhetoric, Irvine here links Wilberforce's skill in rhetoric to his intellectual inferiority, political opportunism, religious immorality, and political danger (the last in that he is pandering to the tasteless mob, which philosophers have often linked to democracy and rhetoric). In describing Wilberforce's speech, Irvine then says that the bishop used "such resources of obvious wit and sarcasm, saying nothing with so much gusto and ingenuity, that he was clearly taking even sober scientists along with him" (6). Just as Plato said about his rival Gorgias, Irvine implies that Wilberforce is a master of deception who prefers performance to substance and style to truth and who poses an immediate danger to the state.

After the bishop, feeling "overcome with success[,] . . . turned with mock politeness to Huxley and 'begged to know, was it through his grandfather or his grandmother that he claimed his descent from a monkey?' " Huxley rose to speak:

[Huxley's] manner . . . was as quiet and grave as the Bishop's had been loud and jovial. He said that he was there *only in the interests of science*, that he had heard nothing to [disprove evolution]. . . . He touched on the Bishop's obvious *ignorance* of the *sciences* involved; explained, *clearly and briefly*, Darwin's leading ideas; and then, in tones even more *grave and quiet*, said that he would not be ashamed to have a monkey for his ancestor; but he would be "ashamed to be connected with a man who used great gifts to *obscure the truth*." (6–7, italics added)

Irvine depicts Huxley as the philosopher/scientist, a man who speaks dispassionately and sincerely, without the misleading color of emotion, in phrases that are crisp and precise; Huxley simply reveals ignorance, exposes rhetoric, and states the truth.[57] In contrasting these two debaters, Irvine is simultaneously describing the dangers and evils of rhetoric, which were exposed unmistakably by a scientist who could see the truth clearly and make it visible for all to see.[58]

It is no coincidence that Irvine later makes his own political claims for evolution, the idea that has demonstrated its truth by overcoming the deceptions of rhetoric: "[Huxley] defended Darwinian evolution because it seemed to constitute, for terrestrial life, *a scientific truth* as significant and far-reaching as Newton's for the stellar universe—more particularly, because it seemed to promise that *human life itself*, by learning *the laws of its being*, might one day become *scientifically rational* and *controlled*" (7, italics added). This passage suggests an issue at the heart of this terminology battle and the recurring conflict between rhetoric and philosophy: Who is going to define *science*? Who is going to control whom? Rhetoric wants to answer this question through a political process, by letting people gather in the *agora* and decide, through a process of persuasion, what they want to do, as based on their interests, values, and beliefs. By contrast, science and philosophy often want this question answered by a few who know the truth, experts who understand the laws of human nature and can therefore control human beings and force them to be rational. Stanley Fish writes: "There is always just beneath the surface of the antirhetorical stance a powerful and corrosive elitism"; at the heart of Plato's *Republic* is also a philosopher-king.[59] The battle between Huxley and Wilberforce, as it has been told here by Irvine, is a move by the knowledgeable elite, a bid for power made in the name of truth by a small group of scientists who feel that their superior rationality entitles them to control what Irvine calls "human life itself."[60]

Interestingly enough, the key details of this story are contradicted by contemporary accounts of the debate. Most accounts were reconstructed from memory by eyewitnesses long after the event actually happened (the most famous account was first recorded in 1898, after an interval of almost forty years); until very recently, many scholars have assumed that actual transcripts could not be found.[61] They apparently passed the story on because they found it interesting rather than because they had ascertained its accuracy.

In 1979, J. R. Lucas published an article entitled "Wilberforce and Huxley: A Legendary Encounter," in which he summarizes all previously known versions and discusses several versions not previously known, including two accounts of the debate written by reporters for British periodicals who were in attendance among the crowd of 700 spectators. Lucas systematically discredits most of the details provided in Irvine's account, concluding that, not Huxley, but Joseph Hooker was the main participant in the meeting who defended the theory of evolution and that Wilberforce focused his attack not on the religious implications of evolution, but on evolution as science. Indeed, some of the scientific questions about evolution asked by Wilberforce in his speech (and in a published review of the *Origin* from which the speech was condensed) were so trenchant that Darwin praised Wilberforce's criticisms in letters to several friends and undertook his first work after the *Origin* "in the areas picked out as weak spots of his theory by Wilberforce."[62]

Lucas does corroborate one of Irvine's details: as part of a chain of reasoning on where the species change could have occurred, Wilberforce apparently did ask whether, according to evolution, it was *his own* grandfather or grandmother who had descended from apes. This statement was then attacked by Huxley, and it was taken by some observers as a lapse in Wilberforce's good manners, primarily because it offended Victorian notions of femininity by applying bestiality to a grandmother rather than a grandfather. One interesting factor influencing perceptions of the debate was, thus, a lapse in Victorian taste.[63]

Irvine's account also contradicts the focus of Wilberforce's position as expressed in his published review. In the review Wilberforce indicates that he does not think it fair to attack evolution on the basis of theology: "We have no sympathy with those who object to any facts or alleged facts in nature, or to any inference logically deduced from them, because they believe them to contradict what it appears to them is taught by Revelation. We think that all such objections savour of a timidity which is really inconsistent with a firm and well-intrusted faith."[64] Wilberforce suggests that evolution cannot destroy real faith and thus avoids making any religious attacks on Darwin's theory. He later adds: "If Mr. Darwin can with the same correctness of reasoning [as Newton used] demonstrate to us our fungular descent, we shall dismiss our pride, and avow, with the characteristic humility of philosophy, our unsuspected cousinship with the mushrooms."[65] Wilberforce did not attack evolution because it contradicted his religious beliefs but because it failed to qualify as science according to his conception. Darwin himself took these criticisms not as dogmatic claims about creationism, but as invitations to explain more clearly and prove more carefully several important aspects of his theory. Wilberforce's review suggests again that this encounter was not primarily a battle between science and religion but rather a battle about the meaning of *science*.

Given these little-known details about this story, what should one make of its astounding popularity? Other explanations have been given besides the one offered here that the story refigures deeply held attitudes toward rhetoric and philosophy. James R. Moore has described the conflict as an institutional battle for power between science and the church. He contends that in response to basic questions about "the identity and character of moral authority in a changing society," Victorian scientists wanted to bring social problems into their own realm, to begin to look for scientific solutions to these problems rather than accepting the solutions proposed by the Church.[66] He ends his essay by describing Darwin's burial in Westminster Abbey (which was planned and accomplished by some of Darwin's scientist-friends, including Huxley). For Moore, this act symbolizes "the Trojan Horse of naturalism entering the fortress of the Church."[67] Moore's interpretation suggests that in this particular terminology battle, science wanted both to define the terms and advance its reputation by making its practitioners into priests and burying them in holy places.[68] This use of religious symbolism by the scientists reflected their conviction that societies are governed better by leaders with scientific knowledge than by those with strong religious beliefs. Both groups wanted their leaders revered, and both recognized the honor implied by burial in Westminster Abbey.

Moore gives another explanation in his book *The Post-Darwinian Controversies*, which attacks the common metaphor of warfare between science and religion. One of Moore's primary targets in this book is Huxley, whose systematic attacks on religion regularly used an explicitly religious rhetoric.[69] Moore also traces printed versions of the Huxley/Wilberforce debate and concludes that one of its chief attractions has been its compelling portrait of science and religion as enemies locked in a death struggle.[70] John Durant concurs with both points: that the Huxley/Wilberforce debate has been "widely seen as a straightforward battle between progressive scientific truth and reactionary theological dogma" and that the "conflict thesis" should be replaced. He indicates that as a result of "increasing dissatisfaction amongst historians . . . there has grown up a school of thought which takes exactly the opposite point of view, namely that, far from having been held back by religious belief, modern science was in fact actively promoted by it."[71] Moore and Durant thus provide another explanation for the popularity of the Huxley/Wilberforce debate. Those who want to emphasize or refight the battle between science and religion could not find better ammunition.

This interesting debate is thus a compelling example of some terminology battles that occur throughout the creation/evolution controversy. In arguing for their position on this fundamental disagreement, some evolutionists have attempted to depict this controversy as a straightforward effort by religious dogmatists to obstruct scientific truth. From the other side (as we shall see in the following chapters), some creationists have made similar efforts to depict evolution as a false and dangerous opinion, which obscures an obvious truth. Both sides have expended great energy in this conceptual and rhetorical labor. The dimensions of their efforts can be glimpsed in the currency and power of the Huxley/Wilberforce debate, which has been elevated into a cultural myth.

Although the battle about the scientific status of creation versus evolution involves particular definitions of terms, I believe that it encodes other fundamental disagreements, especially the larger battle about who shall exercise power and authority in our culture. Should it be the scientists or the priests; the expert opinions of the professionals or the votes of the common people; those who speak in clear and quiet tones or those who use brilliance and bravado? In our Anglo-American culture, built as it is on the assumptions of Enlightenment philosophy, including an implicit distrust of language as rhetoric, it should come as no surprise that many cultural leaders repeatedly argue against the deceptions of rhetoric and for the truth of Enlightenment philosophy.

NOTES

1. Throughout this study the terms *creation* and *evolution* refer to the processes themselves, *creationism* and *evolution* or *the theory of evolution* refer to the differing accounts of these processes, and *creationist* and *evolutionist* refer to proponents of these accounts. Thus, a creationist believes in creationism as an account of how the creation of the universe occurred, and an evolutionist believes in evolution as the name for the process by which life evolved. In general usage, the suffixes -*ist* and -*ism* often convey negative valences, but as used here, these suffixes are not intended to convey such valences for *creationism*, but only to follow standard usages for both terms. In current usage, the terms *creation* or *theory of creation*

are too rare and ambiguous to function as opposites for *evolution* or *theory of evolution*, whereas *evolutionism* seems to undercut the authority of this theory. Thus, even suffixes have rhetorical effects.

2. William Irvine, *Apes, Angels, and Victorians: The Story of Darwin, Huxley, and Evolution* (New York: McGraw-Hill, 1955), 50. Italics added.

3. Later in this chapter, some studies will be summarized that deal with Darwin's persuasiveness and with the Victorian conflict between science and religion.

4. Ernan McMullin, "Introduction: Evolution and Creation," in *Evolution and Creation*, ed. Ernan McMullin (Notre Dame: University of Notre Dame Press, 1985), 1–55.

5. Langdon Gilkey, *Maker of Heaven and Earth: A Study of the Christian Doctrine of Creation* (New York: Doubleday, 1959), 15.

6. The doctrine of biblical inerrancy has played a major role in this controversy but did not explicitly enter the public debates until the rise of fundamentalism in the early twentieth century. The doctrine was first popularized by B. B. Warfield around 1880—somewhat after Darwin's *Origin*. See George M. Marsden, "Understanding Fundamentalist Views of Science," in *Science and Creationism*, ed. Ashley Montagu (New York: Oxford University Press, 1984), 105. The doctrine is discussed at length in Chapter 3.

7. McMullin, "Introduction: Evolution and Creation," 3–8.

8. Peter J. Bowler, "The Changing Meaning of 'Evolution,'" *Journal of the History of Ideas* 36 (1975): 114. This shift is also described in Raymond Williams's entry on *evolution* in his book, *Keywords: A Vocabulary of Culture and Society*, rev. ed. (New York: Oxford University Press, 1983), 120–123.

9. Paul H. Barrett, Donald J. Weinshank, and Timothy T. Gottleber, eds., *A Concordance to Darwin's "Origin of Species," First Edition* (Ithaca, N.Y.: Cornell University Press, 1981). *A Concordance to Charles Darwin's Notebooks, 1836–1844*, ed. Donald J. Weinshank, Stepan J. Ozminski, Paul Ruhlen, and Wilma M. Barrett (Ithaca, N.Y.: Cornell University Press, 1990), indicates that Darwin never used the term during this important eight-year period.

10. *A Concordance to Darwin's "The Descent of Man, and Selection in Relation to the Sexes,"* ed. Paul H. Barrett, Donald J. Weinshank, Paul Ruhlen, and Stephen J. Ozminski (Ithaca, N.Y.: Cornell University Press, 1987), shows that the term *evolution* is used twenty-five times in three forms in this 1871 book.

11. In addition, Spencer emphasized the economic aspects of evolution that later developed into the thesis known as social Darwinism. For Spencer's use of the term *evolution*, see Peggy Rosenthal, *Words and Values: Some Leading Words and Where They Lead Us* (New York: Oxford University Press, 1984), 51–52.

12. Bowler, "The Changing Meanings of 'Evolution,' " 111.

13. Michael J. Buckley, *At the Origins of Modern Atheism* (New Haven: Yale University Press, 1987).

14. Ibid., 85-94 for Descartes, 138–144 for Newton.

15. Ibid., 359.

16. He explains how this conflict leads to atheism as follows: "If an antinomy is posed between nature or human nature and God, . . . this alienation will eventually be resolved in favor of the natural and the human. Any implicit, unspoken enmity between God and creation will issue in atheism" (Buckley, *At the Origins of Modern Atheism*, 363).

17. McMullin, "Introduction: Evolution and Creation," 11.

18. Buckley, *At the Origins of Modern Atheism*, 358–363.

19. William Paley, extract from *Natural Theology*, in *But Is It Science? The Philosophical Question in the Creation/Evolution Controversy*, ed. Michael Ruse (Buffalo, N.Y: Prometheus Press, 1988), 46–49.

20. Paley's is only the best-known example of a group of similar works. Another important work in this tradition was William Whewell's *Bridgewater Treatise* of 1833, from which Darwin took an epigraph for the *Origin* to show how his work fit into this tradition.

21. John Durant, "Darwinism and Divinity: A Century of Debate," in *Darwinism and Divinity: Essays on Evolution and Religious Belief*, ed. John Durant (New York: Basil Blackwell, 1985), 14.

22. Charles Darwin, *On the Origin of Species, 1859*, vol. 15 of *The Works of Charles Darwin*, ed. Paul H. Barrett and R. B. Freeman (London: William Pickering, 1988), 1. Darwin borrowed this phrase from Whewell, whom he called "one of our greatest philosophers." He thus made explicit the connection traced here between his project and natural theology.

23. Neil C. Gillespie, *Charles Darwin and the Problem of Creation* (Chicago: University of Chicago Press, 1979), 14-15, 53.

24. Darwin, *Origin of Species*, 342. Italics added.

25. Gillespie, *Charles Darwin*, 146.

26. David Livingstone, *Darwin's Forgotten Defenders: The Encounter between Evangelical Theology and Evolutionary Thought* (Grand Rapids, Mich.: William B. Eerdmans, 1987); and James R. Moore, *The Post-Darwinian Controversies: A Study of the Protestant Struggle to Come to Terms with Darwin in Great Britain and America, 1870–1900* (Cambridge: Cambridge University Press, 1979). The problem is also discussed in Gillespie, *Charles Darwin*, ch. 5.

27. To complicate the issue yet further, Darwin uses the term *God* in another crucial way in this book: as part of a rhetorical strategy in which he explicitly claims that an evolutionary rather than a biblical account of the origin of species leads to a superior conception of God. Darwin felt that it was better to conceive of a God who works through the orderly laws of natural selection than a God who inefficiently creates each separate species. (Compare Irvine's strategy in the opening quotation for this chapter.) This very shift in the meanings of the term *God* (from the opposite of nature to the creator of order rather than inefficiency) illustrates the rhetoricity of word meanings; here Darwin is using one meaning of *God* to attack another meaning of the same word.

28. Quoted in Robert M. Young, "Darwin's Metaphor: Does Nature Select?" *Monist* 55 (1971): 481. Darwin's personal struggle with the religious implications of evolution has been the subject of several studies. These studies seem to assume that if it could just be determined how evolution affected Darwin's own belief in God, it would be clear whether evolution should undermine other peoples' beliefs. These studies conclude that Darwin remained a theist but rejected Christianity. What to make of this conclusion is the next question; there is no shortage of different answers.

29. The enormous impact of geology on changing attitudes toward the Bible is traced in James R. Moore, "Geologists and Interpreters of Genesis in the Nineteenth Century," in *God and Nature: Historical Essays on the Encounter between Christianity and Science*, ed. David C. Lindberg and Ronald L. Numbers (Berkeley: University of California Press, 1986), 323–350.

30. Moore analyzes the role of Victorian social problems and many other forces in the rise of evolutionism in his essay "1859 and All That: Remaking the Story of Evolution-and-Religion," in *Charles Darwin, 1809–1882: A Centennial Commemorative*, ed. Roger G. Chapman and Cleveland T. Duval (Wellington, New Zealand: Nova Pacifica, 1982), 167–194.

31. Larry Laudan, "The Demise of the Demarcation Problem," in *But Is It Science?* ed. Ruse, 338.

32. Thomas S. Kuhn, *The Structure of Scientific Revolutions*, 2nd ed. (Chicago: University of Chicago Press, 1970), 118–128.

33. Paul Feyerabend, *Against Method*, rev. ed. (London: Verso, 1988); Bruno Latour, *The Pasteurization of France* (Cambridge: Harvard University Press, 1988). The literature on science as persuasion is already extensive and is growing at a rapid rate. Two other excellent examples are Richard Rorty, "Science as Solidarity," in *The Rhetoric of the Human Sciences: Language and Argumentation in Scholarship and Public Affairs*, ed. John S. Nelson, Allan Megill, and Donald N. McCloskey (Madison: University of Wisconsin Press, 1987), 38–52; and Ernan McMullin, "Values in Science," *Philosophy of Science Association* 2 (1982): 1–25.

34. In his introduction to *The Essential Darwin* (London: Allen and Unwin, 1987), Mark Ridley gives some interesting background information (3). Darwin went to Cambridge only after studying medicine unhappily for a time at the University of Edinburgh; he was finally able to persuade his doctor-father that he did not want to become a doctor, too. Darwin was then sent to Cambridge with the expectation that he would become a clergyman. What might have happened had Darwin completed his clerical training rather than gradually switching to science? Such a course of events would have certainly changed the creation/evolution controversy.

35. David L. Hull documents the prevalence of this view in "Charles Darwin and 19th-Century Philosophy of Science," in *Foundations of Scientific Method: The Nineteenth Century*, ed. Ronald N. Giere and Richard S. Westfall (Bloomington: Indiana University Press, 1973), 115.

36. Gillespie, *Charles Darwin* 42.

37. His process of theory/fact formation was actually interrupted and pushed forward by a famous accident of history. Even as he was working on his account, Darwin was sent a manuscript by A. R. Wallace describing virtually the same theory; in response, Darwin felt that he had to move immediately lest he be preempted by Wallace. See Michael Ruse, *The Darwinian Revolution: Science Red in Tooth and Claw* (Chicago: University of Chicago Press, 1979), 160–161. John Angus Campbell adds that Darwin was two-thirds finished with a multivolume version when he received Wallace's letter, whereupon he decided to put it aside. He never finished the work. See "Charles Darwin: Rhetorician of Science" in *Rhetoric of the Human Sciences*, ed. Nelson, Megill, and McCloskey, 70.

38. Young, "Darwin's Metaphor," 469.

39. Charles Darwin, *Calendar of the Letters of Charles Robert Darwin to Asa Gray* (Boston: Historical Records Survey, 1939), 81.

40. Quoted in Gillespie, *Charles Darwin*, 65.

41. Charles Darwin, *More Letters of Charles Darwin*, ed. Francis Darwin, 2 vols. (London: John Murray, 1903), 1:189.

42. Ibid.

43. Hull, "Charles Darwin," 119.

44. Darwin, *More Letters*, 1:189.

45. Young, "Darwin's Metaphor," 445.

46. Ibid., 447, 469.

47. Quoted in ibid., 471.

48. Ibid., 500.

49. Robert M. Young, "Evolutionary Biology and Ideology: Then and Now," *Science Studies* 1 (1971): 177–206. Subsequent pages appear in parentheses in the text.

50. Some ideological implications of the theory have been worked out in studies of social Darwinism as it developed during the late nineteenth century. An article that summarizes previous studies is John C. Greene, "Darwin as a Social Evolutionist," *Journal of the History of Biology* 10 (1977): 1–27.

51. Peter J. Bowler, *The Non-Darwinian Revolution: Reinterpreting a Historical Myth* (Baltimore: Johns Hopkins University Press, 1988).

52. Ibid., 16.

53. Ibid., 69.

54. John Angus Campbell, "The Invisible Rhetorician: Charles Darwin's Third Party Strategy," *Rhetorica* 7 (1989): 55–85.

55. Bowler, *The Non-Darwinian Revolution*, 68.

56. Irvine, *Apes, Angels, and Victorians*, 3. Further page references appear in parentheses in the text.

57. According to several accounts, Huxley, either seriously or jokingly, thought of this opportunity as a religious mission: he exclaimed in a whisper just before he stood up to speak that God had delivered Wilberforce into his hands.

58. In other versions of the story, Huxley used the word *rhetoric* directly in his attack on Wilberforce, accusing him of obscuring the truth through his "aimless rhetoric." See Livingstone, *Darwin's Forgotton Defenders*, 34; and J. Vernon Jensen, "Return to the Wilberforce-Huxley Debate," *British Journal for the History of Science* 21 (1988): 168.

59. Stanley Fish, *Doing What Comes Naturally: Change, Rhetoric, and the Practice of Theory in Literary and Legal Studies* (Durham, N.C.: Duke University Press, 1989), 473.

60. For similar versions of the battle, compare Livingston, *Darwin's Forgotten Defenders*, 33–35; Moore, *The Post-Darwinian Controversies*, 60–62; and Jensen, "Wilberforce-Huxley Debate." Moore gives an overview of many retellings from several countries.

61. On the reconstructed accounts see J. R. Lucas, "Wilberforce and Huxley: A Legendary Encounter," *Historical Journal* 22 (1979): 313.

62. Ibid., 320–321.

63. The point about manners is corroborated by M. J. S. Hodge, "England," in *The Comparative Reception of Darwinism*, ed. Thomas F. Glick (Austin: University of Texas Press, 1974), 3–31. Hodge writes: "Wilberforce's biggest blunder at Oxford was to sabotage his defense . . . by tastelessly and flippantly involving the abhorrent specter of bestial miscegenation in recent human descent. For Huxley and Hooker could seize the opportunity to be 'not amused' and to take the higher moral tone" (11). This effort to correlate the truth of a particular position with the morality of its proponent suggests again that truth is constituted by persuasion, and not vice versa.

In light of the power relations between men and women in Victorian England, it is also no surprise that the point depends on differences between the genders. Much more could be said about this gender issue. In particular, I wonder why male bestiality would be considered more innocuous than the female version.

64. [Samuel Wilberforce], "Review of *On the Origin of Species*," *Quarterly Review* 108 (1860): 256.

65. Ibid., 231.

66. James R. Moore, "1859 and All That," 192.

67. Ibid., 194.

68. The depiction of scientists as priests is treated in more detail by Rorty, "Science as Solidarity," 38. Rorty's article describes the linguistic net around *science* in terms compatible with those developed here.

69. One of the Huxley's most famous attacks is his statement, "Extinguished theologians lie about the cradle of every science as the strangled snakes beside that of Hercules" (quoted in Moore, *Post-Darwinian Controversies*, 60). Huxley's simile provides yet another proof for the connections suggested here; the snake is one of Western civilization's favorite symbols, both for evil and for rhetoric.

70. Ibid., 60–62.

71. Durant, 11.

3

Bryan and the Scopes "Monkey" Trial

In his book, *Six Trials*, about key confrontations between the individual and the state, editor Robert S. Brumbaugh includes an essay on the most celebrated American battle between creation and evolution: the Scopes "Monkey" Trial. In this 1925 trial, John T. Scopes, a high school science teacher, was found guilty of violating a Tennessee law that forbade the teaching of evolution in public schools. According to the author of the essay, Morris Bernard Kaplan, this trial was called, by a Regius Professor of Greek at Oxford, "the most serious setback to civilization in all history."[1] Many other observers have made similar, if less dramatic, judgments about the importance of this trial ever since 1925—in books for various ages and audiences, in history classes, in a Broadway play, even in a 1960 Academy Award–winning film. Why is this trial, which cost Scopes, not his life nor his liberty, but only a $100 fine (paid by someone else), in a conviction that was later overturned on appeal, sometimes described as a threat to civilization itself?

At the end of his essay, Kaplan provides an important clue. He argues that the trial focused not just on one man and his particular teachings about evolution, but on the status of science itself: "The commitment of the state to science requires the recognition of standards that are independent of the political process. . . . These standards are standards of *truth*. . . . Not only is scientific truth beyond the determination of the legislature of Tennessee; in its purest form it is independent of all national and political boundaries."[2] Just as with Darwin's *Origin of Species*, this episode of the creation/evolution controversy also focuses on the relationship between knowledge and politics, or as Kaplan writes, between science as pure truth and the forces that threaten its purity. Earlier in the essay, Kaplan explains that religious fundamentalists got the Tennessee legislature to outlaw evolution in direct contradiction of a state constitutional mandate to "cherish science." He describes the trial as a confrontation between a champion for politics and religion, prosecutor William Jennings Bryan, and a champion for science, defense attorney Clarence Darrow. He suggests a final link in this now familiar conception when he calls Bryan "the master rhetorician" and Darrow "the [soft-spoken] intellectual" (read "philosopher").[3] For Kaplan, this trial threatened civilization because it let rhetoric temporarily overcome philosophy.

The last chapter described earlier episodes of the creation/evolution controversy in part as contests about who should specify contemporary links between key dichotomies, including creation/evolution, religion/science, God/nature, description/explanation, knowledge/belief, truth/error, and rhetoric/philosophy. This chapter analyzes the Scopes Trial as a similar contest about the relationships between such key terms as *science, religion, reason,* and *belief.* When described from the viewpoint of rhetoric rather than philosophy, the Scopes Trial can be seen as a contest about word meanings between competing communities rather than an effort by creationists or evolutionists to misuse important terms and obscure the truth. Most readers are probably familiar with Scopes and Darrow's account of the trial, as retold by Kaplan. It reflects some of the key commitments of people who support science as currently conceived and who reject the values and beliefs of fundamentalist religion.

In order to clarify the conflicting creationist position, this chapter focuses on William Jennings Bryan; it attempts to explain what the key terms of the controversy meant to him, and thus to make sense of his belief in creationism as a function of his worldview. After considering the trial in detail, the chapter concludes by analyzing how Bryan and the trial have been subsequently represented, focusing especially on accounts by Darrow, the journalist H. L. Mencken, the playwrights Jerome Lawrence and Robert E. Lee, and the biologist Stephen Jay Gould. These representations all depict Bryan as an opponent of knowledge, and the trial as a battle between truth and error, reason and belief, and freedom and tyranny, rather than as a contest between different meanings for these key terms. A rhetorical perspective suggests that the Scopes Trial was, finally, not a battle between creationist rhetoric and evolutionist philosophy, but a battle to see whose rhetoric should be used—whose philosophy should prevail.

BRYAN'S KEY TERMS

William Jennings Bryan (1860–1925) has been called, by a recent historian, the most influential Democrat in the United States from 1896 to 1912, and a man of enormous political influence up to his death in 1925.[4] A three-time Democratic nominee for the presidency (in 1896, 1900, and 1908), he was never elected president, but near the end of his life he reported that he had lived to see "nearly every one of the reforms I had advocated written into the unrepealable law of the land."[5]

During his own lifetime, Bryan was seen by many as a very liberal candidate. He resigned in 1915 after serving for two years as Woodrow Wilson's secretary of state because of his strong opposition to World War I. Throughout his career he supported farmers, laborers, and women's suffrage and opposed monopolies, big businesses, and American imperialism. In fact, Bryan made many proposals during his years in power that were seen as truly radical. For example, he proposed that America should never go to war without a popular referendum because "the people who do the dying should also do the deciding," and that money should be drafted as well as men to support World War I.[6] He created such strong opposition among American businessmen that they supported his Republican opponents with massive campaign contributions; the $500,000 contributed in 1896 to McKinley by two supporters alone (J. P. Morgan and Standard Oil) totaled more than the entire Democratic Party fund, and Bryan was outfunded in his 1900 campaign against McKinley by about ten to one.[7] Given the fact that he received more votes in 1896 than any previous candidate in American history in an election with 80 percent voter turnout (though 4.3 percent fewer votes than McKinley), an argument can be made that Bryan lost the presidency because the Republican Party was able to portray him as a substantial threat from the left.[8] Journalist William Allen White summarizes this perception of Bryan by saying that he had "influenced the thinking of the American people more profoundly than any other man of his generation,"[9] and that he "stood for as much of the idea of socialism as the American mind will confess to."[10]

Statements about Bryan's socialism usually come as a surprise to those who first learned about him from accounts of the Scopes Trial. In many of these accounts, Bryan has been depicted as a conservative fundamentalist opposed to evolution and thus, by association, to progress. The leader of the American Progressive movement would probably be surprised to find out that he was considered an antiprogressive. His speeches suggest that he did not consider himself an enemy of progress but rather a friend.

Was Bryan liberal or conservative, progressive or an enemy of progress? This has been an important dilemma for American historians. Earlier historians apparently tried to account for Bryan's opposition to evolution by giving up on his progressivism; they explained Bryan's switch from advocating progressive causes early in his career to his antievolution position late in his career as a retreat toward religious dogmatism or a nostalgic and paranoid reaction to World War I. Such explanations allowed these historians to admire his early progressive politics but to deplore his later antiprogressive politics. However, they have created a dilemma in calling Bryan antiprogressive by separating this important term in the discipline of American history from the man most associated with it. This dilemma was largely resolved by Lawrence W. Levine's 1965 book *Defender of the Faith*, which argued that Bryan supported creationism for exactly the same reasons that he supported his other causes, and thus, that his progressivism was a seamless whole.

This particular debate suggests again that word meanings are rhetorical tools. Those writers who want to call Bryan progressive must decide what to do with his antievolution stance. Was he progressive in his other reforms but reactionary in this one? The problem is that the term *progressive* suggests forward motion, but what counts as forward motion depends on the particular writer. After Bryan began to attack evolution in the early 1920s, many writers wanted, for various reasons, to link evolution with liberal, forward-looking politics and creationism with conservative, backward-looking politics. They argued that Bryan was a conservative rather than a liberal in order to forge links between these terms, and thus to accomplish various kinds of persuasive work in American politics and history.

Bryan himself first expressed tentative opposition to the theory of evolution in his 1904 religious speech, "The Prince of Peace."[11] As a highly sought speaker following his unsuccessful presidential campaigns, Bryan (who was himself a Presbyterian) often spoke on religious topics to audiences predominantly composed of Christians. He delivered this speech, which focuses on the role of Christ as a peacemaker, to many audiences throughout the world ("in Canada, Mexico, Tokyo, Manilla, Bombay, Cairo, and Jerusalem," besides America) (261). Early in the speech Bryan explains that while at college, he went through his own period of skepticism when, as a result of his encounter with evolution, "I became confused about the different theories of creation" (266). On being introduced to the theory of evolution, he interpreted it as a repudiation of the Bible and an attack on Christianity and felt that he must choose between evolution and creation. He resolved this problem as follows: "But I examined these theories and found that they all assumed something to begin with" (266). The realization that all theories begin from assumptions eventually helped him to work out his skepticism: he decided that whereas the theory of evolution assumes "that matter and force existed" on their own and then argues from this assumption that "force working on matter . . . created a universe," he preferred to assume that God had created both the force and the matter. He writes: "We must begin with something—we must start somewhere—and the Christian begins with God" (266). Thus, his opposition to evolution began with his commitment to a Christian God.

During the years between his college decision and this 1904 speech, Bryan examined both creationism and the theory of evolution more closely to decide which one seemed most reasonable to him. In the speech he gives his tentative conclusion: "I do not carry the doctrine of evolution as far as some do; I am not yet convinced that man is a lineal descendent of the lower animals." However, he adds, "I do not mean to find fault with you if you want to accept the theory; all I mean to say is that . . . [I will not be convinced] without more evidence than has yet been produced" (267). In the three additional pages devoted to this topic in this thirty-page speech, Bryan gives several reasons why he does not yet accept evolution, including his sense that it is "a dangerous theory" based on "the law of hate" (267–269). However, before turning to his main topic (Christ as the bringer of hope and peace), he repeats that his audience is welcome to accept evolution if they find it convincing, but he himself is waiting for more proof.

After 1904 Bryan continued to study these issues and gradually solidified his opposition to evolution, ultimately concluding that it was an unproven, irrational, and dangerous belief rather than a scientific fact.[12] Among three famous antievolution speeches, "The Origin of Man" (1924) best summarizes and represents his overall argument. It will be treated at length here to let Bryan make his case in his own words.[13]

"The Origin of Man" begins with two analogies and a set of definitions. In the analogies Bryan compares evolution to an undiscovered planet that is causing "eccentricities in the religious orbit of Christians" and to a poison that is "bringing disorder into the Church" (125). After thus linking evolution to two disturbing natural forces, Bryan quickly, and somewhat cavalierly, defines a hypothesis as a guess, evolution as the hypothesis that all living things developed from one or a few forms of life on earth, atheistic evolution as the hypothesis that life developed by chance, and theistic evolution as the hypothesis that God created living things through evolution. Bryan argues that theistic evolution is worse for believers than atheistic evolution because it removes God from the actual creative process; he calls it "an anesthetic that deadens the Christian's pain while his religion is being removed" (126–129). From the start, the speech thus figures evolution as a disease fatal to Christianity. Even an account of evolution as God's method of creation only deadens the pain.

Bryan argues in the rest of the speech that this disease need not be fatal. Evolution is only a hypothesis or guess; it has not been established as a fact. If people do not accept it as a fact, they will not be susceptible to its negative effects. He presents this position through two steps as indicated in his thesis for the essay: "Before considering the effect of evolution, when accepted as if it were a fact, let us inquire whether it is supported by sufficient evidence to compel a reasonable person to accept it" (129). Bryan bases his inquiry on his own conceptions of evidence and reason. Nowhere in the speech does he argue that creationism must be accepted on faith.

Bryan cites two main types of proof: an absence of evidence for the evolutionary hypothesis (negative proof), and positive evidence that this hypothesis is incorrect. As negative proof he cites the lack of transitional forms between species: "The proof furnished by resemblances [between man and mammals] is completely overthrown by one fact, namely, that it has been entirely impossible to trace any species to any other, notwithstanding the number of species and the resemblances between them" (129). In supporting this claim, he quotes Darwin, Huxley, and a British professor whose lecture he had recently attended, all to the effect that such transitional forms had not been found. In the absence of such evidence, he chides Darwin for speculating about humanity's origins:

> After locating our first parents in Central Africa [in *The Descent of Man*], Darwin asks, "But why speculate?" If he had thought of that in the beginning, he would have been saved the trouble of writing the *Origin of Species* and *The Descent of Man*, both of which are made up of speculations. He used the phrase, "We may well suppose," over and over again, and employed every word in the dictionary that means uncertainty. (132–133)

Bryan takes this admission of uncertainty as proof that Darwin lacked the essential evidence. After critiquing other arguments for evolution, especially the recapitulation argument (the idea that a human fetus goes through all the stages of evolution before developing into a human baby), Bryan concludes that there is "an entire absence of evidence sufficient to support the hypothesis" (137).

Turning to positive proof against evolution, Bryan considers a variety of issues. First, he argues that chemistry itself would have discovered "a pushing force—an internal urge—that tends to lift all matter from lower to higher forms. The fact that chemistry has never discovered the slightest trace or faintest suggestion of such an upward tendency is proof that it is not there—does not exist" (138). Other sciences also disprove evolution, including anatomy: "Anatomy presents convincing evidence that man's body was designed by an Infinite Intelligence and carefully adapted to the work required of him. His eyes, his ears, his heart, his lungs, his stomach, his arteries, his veins, his ducts, his nerves, his muscles—all his parts show that man is not a haphazard development of chance, but a creation, constructed for a purpose" (139). Relying on the design argument (discussed in Chapter 2), Bryan sees the human body and all its parts as proof of planning rather than chance. He does not find rational any explanation that accounts for these parts as a result of "haphazard development," and writes that in such cases, "the evolutionists not only reason without facts, but they reason ridiculously" (139). Then he gives several examples chosen from contemporary sources to show how unpersuasive and irrational these explanations seem to him.

How did evolutionists at that time account for the eyes? "A piece of pigment or a freckle on one spot of the skin" that appeared one day by accident was irritated by the increased heat of the sun until it grew its own nerve. "Then another freckle and another eye" grew in exactly the right place so that there could be two eyes (140). What was an evolutionary explanation for the legs? "A little animal one day discovered a wart on its belly" that it found useful for locomotion; it used this wart "until it developed into a leg. And then another wart, and another leg" (141–142). How did evolutionists account for the fact that people awake during dreams of falling? According to a Pennsylvania college teacher, "We dream of falling because our ancestors fell out of trees fifty thousand years ago; but, he says, we never dream of being hurt when we fall—his explanation being that those who fell and were killed had no descendants, and that we must, therefore, have descended from those who fell and were not killed" (142). After glossing these explanations, which were chosen because of their weaknesses, Bryan asks: "Is it necessary to believe all this tomfoolery at the risk of being called ignorant if we reject it?" (143). To him it was far more rational to believe in creationism.

He ends his inquiry into the evidence for evolution by simplistically relating his key terms, *evidence* and *hypothesis*, to two other crucial terms, *fact* and *truth*:

> Christianity does not fear any *truth* that science has discovered or may discover. All truth is of God, whether it is revealed in the Bible or by nature;

therefore, truths cannot conflict. It is not truth that Christians object to; they object to *guesses* put forward without verification and substituted for "Thus saith the Lord." Newton's definition of the law of gravitation deals with a *fact*, and that fact has done Christianity no harm. It does not contravene a single Bible truth. So with the roundness of the earth; it is a fact, and provable, and it does not disturb Christianity. But evolution is not a fact; it is not provable; it is merely a guess. (143–144)

This statement reveals Bryan's commitment to a pre-Darwinian conception of science as the other "book" besides the Bible. Lacking much knowledge of science himself, he is convinced that confirmed scientific hypotheses can be easily proven and do not conflict with the Bible, which is the other source of truth. By these criteria, Bryan concludes that evolution is not a fact of science; it is a guess, which cannot be proven; a statement that lacks authority; a contradiction of the Bible. Bryan uses this linguistic net within his own easily attacked conception of science to argue that he is not rejecting, but rather defending, the truth of creationism against the error of evolution.

Having argued in the first part of the essay that evolution need not be accepted by reasonable persons, Bryan turns in the second part to a discussion of its negative effects. He explains this hinge in his argument as follows: "The objection to evolution, however, as an explanation of life, is not primarily that it is not true—many things that are false are scarcely deserving of attention. Neither is the ridiculousness of the explanations of evolution the chief reason for rejecting it. . . . The principal objection to evolution is that it is highly harmful to those who accept it and attempt to conform their thought to it" (144). The main danger is that evolution tends to destroy faith in the Bible and in God. As proof, Bryan cites Darwin's own loss of Christian faith. Then he cites the thought of Friedrich Nietzsche, which he sees as the natural endpoint of evolution, to show that evolutionists conclude that God is dead and that might makes right. Finally, he cites personal experiences and a study by a Bryn Mawr College professor to prove that evolution has caused many college students to lose their Christian faith. Bryan ends this section by asking in a biblical allusion: "What shall it profit a student, boy or girl, if he gain an education and lose his soul?" (149) Bryan believes that evolution leads to a loss of faith, and possibly even damnation, outcomes that, he assumed, his mostly Christian audience would find disastrous.

In the final section of this speech, Bryan asks "What is the remedy?" He gives three specific answers that all grow out of his conception of representation in a democracy. First, he suggests that every Christian congregation openly discuss creationism and evolution: "Let each church member state his or her position candidly and honestly, leaving the majority to decide what the position of the church shall be" (152). He explains his position as follows:

> Freedom of conscience is guaranteed in this country and the guarantee should never be weakened. Freedom of speech is also guaranteed, and no restrictions on it should be permitted. The individual has the right to think for himself, to believe what he likes, and to express himself as he pleases. But freedom of conscience and freedom of speech are individual rights and belong only to individuals, as individuals. The moment one takes a representative character, he becomes obligated to represent faithfully and loyally those who have commissioned him to represent them. (152)

For Bryan individual freedom of speech and conscience stop whenever one becomes a representative of a group rather than a spokesperson for oneself. Thus, he encourages each church to determine its own stance by vote of its members.

His second proposal is, "Stop the teaching of evolution—not as a mere hypothesis, but as a fact—in church schools" (153). Turning in his final proposal to public schools, he states:

> Likewise with public schools; teachers in public schools must teach what the taxpayers desire taught—the hand that writes the paycheck rules the schools. A scientific soviet is attempting to dictate what shall be taught in our schools and, in so doing, is attempting to mould the religion of the nation. . . . These scientists [who he says numbered about 1 per 10,000 in the general population] are *undermining the Bible* by teaching daily *that which cannot be true if the Bible is true*. These assaults upon the Bible are not based upon *established facts* or *demonstrated truths*, but, as has been shown, are built upon cobweb theories as unsubstantial as the "fabric of a dream." (154–155, italics added)

This quotation indicates the importance Bryan places on the key term *truth*, which he links to *facts* and *demonstrations* in connection to the Bible itself, but nevertheless sees as a function of a democracy. Bryan holds that if evolution is true, the Bible must be false. He argues that the Bible should not be attacked in schools until evolution has been proven and a majority in the community has been persuaded to believe it. Those evolutionists who want to teach it in public schools without such "established facts" are unfairly attempting to advance their own beliefs, "to mould the religion of the nation."

In the next paragraph, he comments more explicitly on a biology teacher's freedom of speech:

> If a teacher of evolution insists that he should be permitted to teach whatever he pleases, regardless of the wishes of the taxpayers, the answer is obvious. He should teach what he is employed to teach, just as a painter uses the colours that his employer desires, just as the army or navy officer uses the equipment provided by the government and directs it against those whom the government desires attacked; just as the public official carries out the will of his constituents. Would a teacher be permitted to teach in any public school in the United States that a monarchy is superior to a government in which the people rule, or to advise pupils that they should not obey the law? (155)

In highly questionable analogies comparing a teacher to a painter, a soldier, and a public official, Bryan argues that all these employees ought to do what their employers want. By acting as representatives of the community that hired them, teachers forfeit their right to free speech as individuals and should not teach evolution, just as they should not attack democracy nor undermine the law. He then makes more explicit the relationship between democracy and individual freedom:

> We do not interfere with freedom of conscience or with freedom of speech when we refuse to pay a man for teaching things that we think are injurious, especially to the young. Christians are required to build their own colleges in which to teach Christianity; why should not atheists be required to build their own colleges in which to teach atheism? . . . Why should a few people demand pay from the public for teaching a scientific interpretation of the Bible when teachers in public institutions are not permitted to teach the orthodox interpretation of the Bible? By what logic can the minority demand privileges that are denied to the majority? (156–157)

Bryan contends that freedom of speech does not apply when an unproven hypothesis such as evolution is taught to young children in a public school. Even the Genesis account cannot be taught in the public schools; it is only taught in Christian schools. To be logically consistent, the evolutionists ought to build their own schools to teach their own unproven hypothesis, just as Christian groups must build their own schools to teach creationism. For Bryan, the only logical policy is the exclusion of both.

Bryan concludes his speech by underlining the dangers of evolution, which, he claims, is "the greatest menace to civilization as well as to religion" because "belief in God is the fundamental fact in society; upon it rest all the controlling influences of life. Anything that weakens man's faith in

God imperils the future of the race" (157). Within the framework of his primary commitment to God and his conceptions of science and democratic representation, Bryan saw evolution as an irrational and dangerous belief and creationism as a position supported by good evidence and good reasons.

In response to a request like Bryan's for solid proof of the theory of evolution, contemporary evolutionists can begin to glimpse the difficulty of their task. To persuade a creationist like Bryan, evolutionists must begin to explain what they mean by *proof*, to indicate why evolution cannot be proven by something as simple as an apple falling to the earth or a trip around the world. If they succeed at this task, they can then begin to explain what proofs *do* exist for evolution, thus helping the creationist understand why these *are* proofs and *what* they actually prove. However, this approach is fraught with problems. First, the creationist may see this entire effort as a deceptive shift of the word *proof*. Moreover, even if he or she listens, all the explanations will likely be taken as a request to believe the evolutionist, which, of course, was the very thing the creationist was trying to avoid when first asking for proof. The statement, "Prove it!" is seldom a request to negotiate about what counts as proof.

If the conversation continues, the evolutionist will probably bring up fossils as the primary proof that species have changed throughout the long history of the earth.[14] However, the age of the earth itself will probably also require proof, since creationists typically interpret Genesis to mean that the earth is relatively young. What could prove that the earth is very old? Carbon dating and geological processes are used, but what do they prove, and how? Each new proof creates the same problems. Every proof of evolution is not a simple correlation between fact and theory, but an interpretation: the facts cited as proof are only facts within a framework of other facts, assumptions, and perceptions. There is no solid ground from which to begin. Similar interpretive difficulties would be encountered by a creationist attempting to prove creationism to an evolutionist; the creationist would probably use the Bible as a source of proof rather than fossils and interpret fossils as proof of something else. Both sides agree that one needs proof, but the word *proof* itself does not point to the same thing on both sides. Thus, the controversy turns in part on who gets to define *proof*.

As has been shown at length in "The Origin of Man," Bryan's attack on evolution depends on a simplistic conception of science and a strict view of the duties of a representative in a democracy, but it nevertheless makes extensive use of such terms as *reason* and *proof* in an internally consistent way. In other speeches Bryan makes similar use of the words *education* and *science*.

In a commencement speech entitled "Man" and delivered at Nebraska State University in 1905 and later at Illinois College, Bryan states:

> There is in some quarters a disposition to regard what is contemptuously called "book-learning" as of little value except in the professions. No error can be more harmful. . . . Whether a boy intends to dig ditches, follow the plow, lay brick upon brick, join timber to timber, devote himself to merchandising, enter a profession, engage in teaching, expound the Scriptures, or in some other honorable way make his contribution to society, I am anxious that he shall have all the education that our schools can furnish.[15]

Bryan adds that a person who seems adversely affected by education has a problem that "can be traced to a deficit in purpose rather than a surplus of learning."[16] He shows his support for science as he conceives it in a speech given at the Scopes Trial, where he indicated that he was a member of the American Academy for the Advancement of Sciences; in "The Bible and Its Enemies," where he argues that "science can do anything when it builds on facts";[17] and in his post–Scopes Trial speech, where he writes: "Give science a fact . . . [and] it is not only invincible, but it is of incalculable service to man."[18] In "The Menace of Darwinism," he concludes: "Science has rendered invaluable service to society; her achievements are innumerable—and the hypotheses of scientists should be considered with an open mind. Their theories should be carefully examined and their arguments fairly weighed, but the scientist cannot compel acceptance of any argument he advances, except as, judged upon its merits, it is convincing."[19] The problem with evolution from Bryan's perspective was that it had not yet attained the status of a fact; indeed, the evolutionists seemed to him a group of scientists unwilling or unable to persuade the public to agree with them. Instead, they were ready "to force irreligion upon the children under the guise of teaching science" (322).

Why would scientists use their power in this way? Agreeing with Irvine's claims about science (as discussed in relation to the Huxley/Wilberforce debate), Bryan argues that they desire to control other people. He writes about the evolutionary hypothesis: "Its only program for man is scientific breeding, a system under which a few supposedly superior intellects, self-appointed, would direct the mating and movements of the mass of mankind" (334). Generalizing from evolution, Bryan expresses his reservations about science as a whole:

> The greatest danger menacing our civilization is the abuse of the achievements of science. Mastery over the forces of nature has endowed the

twentieth-century man with a power which he is not fit to exercise. Unless
the development of morality catches up with the development of technique,
humanity is bound to destroy itself. . . . Science is a magnificent material
force, but it is not a teacher of morals. It can perfect machinery, but it adds no
moral restraint to protect society from the misuse of the machine. (337–338)

Bryan saw science as an inventor of powerful machines. He argued that it
must be controlled by persuasion; it must convince the public of the value of
these machines and protect it from their misuse. He feared the power of science when it included coercion to advance its goals. He opposed evolution
because he saw its teaching in public schools as a coercive attempt by scientists to unleash a destructive power and weaken society's moral restraints.
His opposition makes sense if one accepts his definitions of terms and shares
his worldview, based as it is on commitments to a Christian God, a literal Bible, and a strict democratic process that attempts to eliminate dangers to society as a whole. The challenge of convincing someone like Bryan to see
things differently is that he did use the key terms of this debate in what he
and many others saw as a consistent way. Evolutionists of his era could not
just argue for the opposite terms; they had to decide how to define the terms
differently so as to reverse the polarities and linkages that were woven into
the arguments Bryan made.

THE SCOPES TRIAL

Not only did Bryan defend his position vigorously during the Scopes
Trial, but his influence led to passage of the law under which Scopes was
tried. During the years before the trial, Bryan traveled the country, gathering evidence against evolution and speaking out against it.[20] George Marsden notes that, contrary to the popular conception of fundamentalists as
anti-intellectuals, their theology assigns "vast importance to ideas."[21] Bryan
clearly emphasized the importance of this particular idea, by linking it to
three of his most important causes: his opposition to World War I, his defense of Protestant Christianity, and his criticism of capitalists, whom he
felt had created vast social problems in America. Tennessee was the third
state to pass an antievolution statute, after Oklahoma and Florida. In Oklahoma, evolution had been outlawed in a bill that simultaneously provided
for free elementary school textbooks. In Florida, Bryan had personally written letters to state legislators in which he requested (successfully) that the
law apply only to teaching evolution "as true."[22] As part of his antievolution
campaign, Bryan gave a speech in 1924 in Nashville; this speech was repro-

duced and put on the desk of every Tennessee state legislator, where it aided in the passage of the Butler Act in 1925.[23]

In brief, the Butler Act declared it illegal for any teacher in a public school to teach "any theory that denies the story of the divine creation of man as taught in the Bible, and to teach instead that man is descended from a lower order of animals," and it set the penalty for violations at a fine between $100 to $500. Bryan himself was opposed to any penalty clause for violations; the Florida law lacked such a clause because he felt the state should simply declare its policy, thus letting the teachers know what their employer expected. However, the act, complete with the penalty clause, was signed into law by Governor Austin Peay on March 23, 1925, which brought him ridicule in the national press and from many other opponents, both within and outside Tennessee.[24]

Not long after the Butler Act was passed, the American Civil Liberties Union, which was still a fledgling organization, having been founded during World War I (in New York City), advertised in various Tennessee papers for a volunteer to act as defendant in a case testing the constitutionality of the new law. After an informal business meeting in the local drugstore, the town leaders of Dayton, Tennessee decided that such a test case would provide excellent publicity and improve the town's business. They talked Scopes into offering himself as the defendant, even though he was actually a coach at the high school who thought he had only taught Darwin once as a substitute teacher in preparing the students for a biology exam (and was uncertain later whether he had actually ever taught the subject). When the case came to Bryan's attention, the "Great Commoner" decided to offer his services to the prosecution because he felt the trial would provide an excellent platform from which to attack evolution. Subsequently, Clarence Darrow, a famous trial and labor lawyer, heard about Bryan's involvement and decided to offer his services to the defense, in part because (he confessed later), "For years [I'd] wanted to put Bryan in his place as a bigot."[25] The more one knows about the events that led to the trial, the less it sounds like a solemn confrontation between truth and error; indeed, it was the first modern-style media event ever staged in the United States. Dayton's community leaders even took extra precautions to keep the trial in town after they found out that Chattanooga was trying to steal it and thus benefit from the free publicity.

The trial itself took place from July 10 to July 21, 1925, during eight days of proceedings before thousands of spectators and hundreds of reporters.[26] Not only was the courtroom packed, but loudspeakers had been set up outside on the lawn to broadcast the proceedings, and the trial was actually moved outside when the courtroom became too hot.

Briefly, the course of the trial can be outlined as follows: The prosecution decided that it would limit its case to proving that Scopes had violated the law by using a biology textbook to teach evolution as a fact. It claimed that the legislature had the right to regulate the public schools so as to exclude dangerous beliefs according to the will of the majority. In other words, its case paralleled Bryan's position. The defense decided to use the opposite strategy: it brought in a group of expert witnesses to testify that evolution had been proven scientifically and that there was no necessary contradiction between evolution and creation as these terms were properly understood. The defense also tried to show that the law violated the constitutional separation of church and state and interfered with free education and the freedom of speech. They made the latter arguments during pretrial motions, but these arguments were rejected by the judge, who ruled that the law was constitutional.[27] After the prosecution had called its witnesses and rested its case, the defense turned to the first group of arguments and called its first expert witness. At this point the prosecution objected to testimony by any expert witnesses, contending that the defense was trying to prove that the law should never have been enacted in the first place rather than that it had not been violated by Scopes. In a stunning defeat for the defense, the court upheld this objection as well; the expert scientists who had gathered in Dayton from around the country were only allowed to introduce into the record written statements to be used on appeal. Having little other basis for his defense of Scopes except attacks on the wording of the law, Darrow decided to call Bryan to the stand as an expert witness on the Bible in a cross-examination that has become legendary.

At the conclusion of the trial, the jury found Scopes guilty after only eight minutes of deliberation, and the judge sentenced him to the $100 minimum fine. However, the conviction was overturned on appeal on a technicality: the fine had been levied by the judge rather than the jury, as required by state law. As a result, the defense was not able to appeal the conviction as planned (they had hoped to take the case all the way to the U.S. Supreme Court), and the Tennessee law remained in force until its repeal in 1967.

Throughout the trial, the defense used the same key terms for its arguments as Bryan did, but in reverse. Whereas Bryan and the other prosecutors held that evolution contradicted creation and should not be taught because it was an unproven and dangerous belief, Darrow and the other defense attorneys argued that evolution was a scientific truth and, moreover, that evolution and creation are not necessarily opposed. The defense thus attempted to weave a linguistic net in which evolution was linked to science, knowledge, and freedom; and creation was linked to religion, ignorance, and intolerance. By attacking Bryan's rationality and pointing out the dangers of the law, they hoped to persuade the jury (and, more important, the world) that any opposition to evolution was a dangerous opposition to truth, science, education, and freedom.

The strategy was clear in Clarence Darrow's longest speech to the court, which he made on the second day of the trial.[28] Darrow began his speech by calling the law itself "a foolish, mischievous, and wicked act," for which Bryan was responsible. Then Darrow announced that he would argue the case "as if it were a death struggle between two civilizations" (74). These two civilizations were a society in which "people believed in freedom, and . . . no men felt so sure of their own sophistry that they were willing to send a man to jail who did not believe them" and a society that upheld "bigotry and ignorance" in "as brazen and as bold an attempt to destroy learning as was ever made in the Middle Ages" (75). By the end of Darrow's first two paragraphs, this trial had been defined as a contest between freedom and restriction, enlightenment and sophistry, tolerance and bigotry, knowledge and ignorance, the modern world and the Middle Ages, the age of reason and the age of belief.

How could one recognize the participants on each side of this conflict? Shifting his argument to the present tense, Darrow writes: "Intelligent, scholarly Christians . . . by the millions in the United States find no inconsistency between evolution and religion"; only "a narrow, ignorant, bigoted shrew of religion" perceives a conflict between the two theories (76). The members of this latter group (the creationists represented by Bryan), are not only unreasonable in their views of creation and evolution, but "are after everybody that thinks" (79) and "neither know nor care what science is except to grab it by the throat and throttle it to death" (83). Throughout his speech Darrow portrays his opponents in the trial as enemies of freedom, learning, and science because they believe in creationism rather than accepting evolution as a fact. If these people have their way and "ignorance and bigotry" are "permitted to overwhelm . . . the rights of man," Darrow writes, then "we [will be] taken in [such] a sea of blood and ruin that all the miseries and tortures and carrion of the Middle Ages would be as nothing" (83). He concludes the speech as follows:

Today it is the public school teachers, tomorrow the private. The next day the preachers and the lecturers, the magazines, the books, the newspapers. After a while, your honor, it is the setting of man against man and creed against creed until with flying banners and beating drums we are marching backward to the glorious ages of the sixteenth century when bigots lighted fagots to burn the men who dared to bring any intelligence and enlightenment and culture to the human mind. (87)

Darrow's argument depends on taking creationism as a dangerous belief (he uses the term *creed*) that is opposed to the scientific truth known as evolution, which brings "intelligence and enlightenment and culture to the human mind." Using his own kind of overstatement and bravado, he attacks the creationist position as strictly irrational; it is clearly a threat to many of his crucial values and beliefs, including freedom of thought and religion.

Of course Darrow's position depends on the claim that it is indeed irrational to reject the theory of evolution; that in seeing a conflict between creation and evolution, the creationists are ignoring obvious facts and opposing freedom, education, science—even thought itself. However, this position depends on accepting his criteria for rationality and his meanings for these terms. In fact, Bryan used the same terms to support his creationist position; he argued exactly the opposite of Darrow's position—that evolution is not scientific because it has not been proven by obvious facts, that it is not rational to a thinking person, and that to teach it in the public schools is to abuse freedom and education. When Bryan finally rose, on the fifth day, to make his long trial speech against evolution, he did not defend ignorance, attack freedom or reason, oppose education, or glorify the persecution of heretics. He argued that evolution is not science and that it has been taught unfairly in public schools because it is itself what Darrow would call a "dangerous creed." Their conflict involves a terminology battle about what counts as a belief and what as a scientific fact, against the background of differing conceptions of the greatest intellectual dangers. For Darrow, the great enemy is religious persecution; for Bryan, the loss of faith in God. They are speaking at cross-purposes and out of different worldviews, yet both are using the same key words in an attempt to convince the audience to accept their positions.

In this speech, Bryan also mentioned the other crucial issue for the trial, and indeed for the controversy as a whole: the issue of biblical inerrancy. This was a major battleground for the competing notions of rationality that were crucial to this trial. Whereas creationists still see inerrancy as a crucial basis for rationality, evolutionists see people who accept this belief as irrational, and consequently they attempt to undermine their credibility and dismiss their definitions of terms.

As discussed in Chapter 2, a foremost question in this controversy is how to define the key terms, *creation* and *evolution*. If these terms are opposites, then a choice must be made between them and the dominant term will tend to erase its opposite. However, if the terms can be reconciled, they will tend to collapse into each other unless their difference can be asserted by relating them to another dichotomy. Bryan and the Tennessee creationists saw the terms as irreconcilable opposites, whereas Darrow and the evolutionists thought that the two could be reconciled by making evolution God's method of creation. In arguing against the creationists' meanings for *reason* and *belief*, the defense decided to attack their position directly on this question of reconcilability. Bryan gave them a target in this quotation from his trial speech: "In this state they cannot teach the Bible. . . . We will not teach the Bible. . . . The question is can a minority in this state come in and compel a teacher to teach that *the Bible is not true* and make the parents of the children pay the expenses of the teacher to tell their children what these people believe is *false* and *dangerous?*" (172) Agreeing with Bryan's oppositions between truth and falsehood and his application of this opposition to the Bible itself, many creationists since Bryan have opposed evolution because they think it "teaches that the Bible is not true." They hold that the biblical account of creation and the theory of evolution are opposites; because the Bible is true, evolution must be false. Although the defense chose to argue instead that the opposition does not exist, many evolutionists agree with this logic and use it to reach the opposite conclusion: that evolution is true, and therefore the Bible must be false.

This notion of irreconcilable conflict between creation and evolution depends on the concept of biblical inerrancy, which developed late in the nineteenth century in the United States as an important theological position and a historical key to American fundamentalism.[29] James Barr defines biblical inerrancy as the belief that the Bible is free of error of any kind. He writes: "The inerrancy of the Bible, the entire Bible including its details, is indeed the constant principle of rationality within fundamentalism."[30] This position on the Bible grounds all fundamentalist arguments; it is the measure—albeit a very narrow one, which is fraught with many disturbing implications for nonfundamentalists—of reasonableness itself. In evaluating a statement for its truth, fundamentalists compare the statement to the Bible, resolving any conflict between the two by rejecting the statement and keeping the Bible.

In the case of the Genesis account of creation, such an approach creates an obvious conflict. Genesis reports that God made the world in six days, made each creature after its own kind, and made man from the dust of the earth. Darwin's books argue that the world is millions of years old and that new species of living things, including humans have evolved from old species. How do creationists resolve this conflict? Because they cannot find an acceptable way to reconcile the apparent contradictions, they choose to reject evolution and retain the Genesis account. This is the only rational course available to them unless they give up their central belief in an error-free Bible or find another way to interpret the biblical account. Using the same logic, many evolutionists conclude that the Bible contains errors and thus reject it in favor of evolution.

Both the creationist position on inerrancy and the evolutionist rejection of the Bible typically depend on a strict notion of the purity and transparency of language (a notion that was described in Chapter 1 as a positivist account of word meanings). George Marsden indicates that fundamentalists see the Bible as a "book dropped out of heaven" and view "their own interpretations as simple and straightforward interpretations of fact according to the plain laws of common sense and the common man."[31] Using the same account of words as names for obvious things, some evolutionists hold that their theory is supported by plain facts that are interpreted simply and straightforwardly; they differ from the creationists primarily in thinking that their account has been deduced from natural facts rather than descending from God. How do these senses of biblical inerrancy and natural fact play out in Bryan's creationism and Darrow's support for evolution?

Bryan's creationism is, in part, an effort to link the term *true* to the Bible. Throughout his years of opposing evolution, Bryan insisted hundreds of times that the Bible and its account of creation were true and the theory of evolution was false. As a Protestant who was committed to the Bible, Bryan felt that to accept the truth of the Bible was to see it as the word of God, an important proof of God's existence, and the only means to salvation. If the Bible or any part of it including the creation account were to be proven false, then, according to the strict logic of Bryan's dichotomy, the Bible itself could only be false. However, to prove the Bible false would be to disprove the existence of God and to undermine any belief in Christianity. Thus, for Bryan, this issue was the crux of the entire controversy.

By *proof* for the Bible, fundamentalists following Bryan do not mean scientific evidence; indeed, they do not doubt that the Bible is true until they see proof in nature. Rather, their proof might include their experiences with the Bible, statements they have heard from other people who also call the Bible true, beliefs they have been taught by their parents, and so forth. By saying that the Bible is *true*, fundamentalists mean such things as that they believe it came from God, they trust its statements, and they think it describes real people and events. Because the Bible is true for them, they take its statements about subjects such as creation as a rational basis from which to evaluate the truth of other statements. If a statement contradicts the Bible, that statement must be false.

What Darrow tried to do after his scientific witnesses had been barred from testifying was to insist that the Bible could not be true in the sense demanded by biblical inerrancy. Darrow and the other defense attorneys claimed that they were perfectly willing to let the Bible be true in some other sense, but not in claiming that God created humanity and the world as described in Genesis. The problem with this attempt from the creationists' point of view was that it undermined the one meaning of *truth* most vital to them—the sense in which the Bible is *true* because it is *inerrant*. To admit that the Bible contained an error about creation was to consider it false, and thus to disbelieve it. By this complex chain of linguistic associations and simplifications, the controversy came to turn on biblical inerrancy.

The defense made their case most clearly in a speech given by Dudley Field Malone, another defense attorney and Bryan's former undersecretary of state.[32] Malone's speech, which he gave soon after Bryan's speech on the fifth day, begins by responding to Bryan's defense of the Bible: "I know Mr. Bryan. . . . I know this, that he does believe—and Mr. Bryan, your honor, is not the only one who believes in the Bible" (183). Malone then argues that even among believers like Bryan, there are many different interpretations of the Bible and many accounts of creation.[33] The problem occurs (as it did in the case of Galileo, which he narrates as an example) when someone takes "a literal construction of the Bible [as] truth, which is revealed" (183). Picking up on the term *revealed*, Malone begins to distinguish between religious and scientific truth: "Your honor, there is a difference between theological and scientific men. . . . The difference between *the theological mind* and *the scientific mind* is that the theological mind is *closed*, because [for it the truth] is revealed and is settled. But the scientist says no, the Bible is the book of revealed religion, with rules of conduct and with aspirations—that is the Bible" (184). Asserting a difference between the truths possessed by science and theology, Malone links this difference to the opposition of *open/closed*. He illustrates the difference by suggesting that the scientist refuses to limit truth, to remain trapped inside a closed mind like the theologian; the open-minded scientist goes beyond the enclosure by seeing the Bible only as a book of "revealed religion." Thus, the scientist ventures beyond this dull and settled space of the theologian.

Malone clarifies exactly what remains within this enclosure later on (after he tells another vivid story about the Moslems who burned the library at Alexandria because they felt the Koran contained all truth):

> But these gentlemen [the Scopes prosecutors] say the Bible contains the *truth*—if the world of *science* can produce any truth or fact not in the Bible as we understand it, then destroy science but keep our Bible. And we say "Keep your Bible." Keep it as your consolation, keep it as your guide, but keep it where it belongs, in the world of your own conscience, in the world of your individual judgment, . . . keep your Bible in the world of *theology* where it belongs and do not try to tell an intelligent world and the intelligence of this country that these books written by men who knew none of the accepted *fundamental facts of science* can be put into a course of science. (185)

Like Darrow, Malone depicts Bryan and the creationists not as people with different conceptions of science and truth, but as people willing to destroy the truth of science to preserve the truth of their book. In contrast to their opposition to scientific truth, he is willing to keep their Bible, to retain both types of truth—but note the role he leaves for the Bible: it can be kept only if it is limited to a private space and used by an isolated individual. It can remain as a source of consolation and guidance only; it must remain metaphorically cloistered, just like the theologian. Specifically, the Bible is forbidden to appear in the public space, in "an intelligent world" and an "intelligent" country, and especially in science classes, where knowledge is taught rather than religious beliefs. Malone's argument does not make room for two equal truths, but implicitly separates the greater truth from the lesser. Bryan and the creationists were probably not relieved to hear that they could keep their Bible as devalued by Malone.

Malone concludes his speech by asking the court to implement this difference between truth claims. After blaming Bryan for expressing dangerous opinions on a subject about which he is unqualified to speak (and thus repeating again the defense's attack on his credibility as a speaker), Malone asks the court to protect the children from this biblical danger: "For God's sake let the children have their minds kept open—close no doors to their knowledge; shut no door from them. Make the distinction between theology and science. Let them have both be taught. Let them both live" (187). In this quotation, Malone appeals to God against the threat of enclosure made by theology. Then he curiously asks that theology (a subject that, by the end of his speech, he hardly seems to admire) also be allowed to live. How does one guard against the threat from theology without murdering the threatener? One must draw a distinction between fields that deliver different types of truth (but only one of which delivers real truth). He ends his speech as follows:

> There is never a duel with truth. The truth always wins and we are not afraid of it. . . . We are ready to tell the truth as we understand it and we do not fear all the truth that they can present as facts. . . . We feel we stand with progress. We feel we stand with science. We feel we stand with intelligence. We feel we stand with fundamental freedom in America. We are not afraid. Where is the fear? We meet it, where is the fear? We defy it. (187–188)

Using the same list of positive terms used by Darrow (*progress, science, intelligence, freedom*), in the same style of overstatement and bravado, Malone links them all to the free public teaching of evolution. He wants the truth (or at least that part of it that can be presented "as facts") to go forth (although he says it will do so on its own); he defies the fear that he feels must underlie any opposition to evolution.

What Malone's speech does not acknowledge is that these same terms can be used by people who support the opposite position. Creationists think of the Bible as true in something other than Malone's limited sense. They feel that the danger arises not from their own intolerance, but from evolutionist unbelief. They contend that facts come from the Bible rather than the scientific method. Creationists do not perceive their belief in the Bible as a decision to remain within a narrow enclosure; nor do they want to limit the Bible to a private rather than the public role or agree that to transgress this limit is to act against intelligence. They hold that to limit the truth of the Bible is to deny its truth overall. Thus, the creationists would dispute all Malone's linguistic linkages.

In his speech Malone thus defends evolution by using terms and images that claim to have obvious meanings but that do not mean the same thing to everyone nor reflect values held by all. From their own perspective, the creationists feel they are being rational and, more important, defending the truth of the Bible against unbelievers and science against unproven hypotheses. To call them irrational is not to give a fact about their intelligence, but to disagree about standards of rationality. Malone does not reveal correct word meaning from some objective position but rather from a particular position that defines rationality in terms of such notions as scientific evidence, linguistic coherence, and tentative doubt. By contrast, the creationists define rationality in terms of conformity with the Bible. To accept Malone's distinctions is to agree with his values and to deal with the Bible on the basis of reason more than belief, doubt more than faith, and tolerance for different interpretations more than certainty about one. Both conceptions rely on different key values, as is evident when evolutionists call creationists irrational and creationists call evolutionists atheists and unbelievers.

This conflict between values and the terms used to describe them is nowhere clearer than in the most famous part of the Scopes Trial, Darrow's cross-examination of Bryan as an expert witness on the Bible. It was highly unusual for the defense to call Bryan at all, given that the court had banned expert testimony and that Bryan was part of the prosecution team. At first the other prosecuting attorneys objected, but Bryan indicated that he was willing to be questioned on the condition that he could also question Darrow. During two and a half hours of questioning on the seventh day of the trial, Darrow relentlessly pressed Bryan on biblical inerrancy. In trying to demonstrate the inconsistency of Bryan's answers, Darrow attempted to ridicule his belief in an inerrant Bible, to portray creationism as irrational, and thus to uphold the terms and key issues as defined by the defense.

Darrow started the cross-examination by verifying that Bryan had studied the Bible at length. Then he asked a crucial question: "Do you claim that everything in the Bible should be literally interpreted?" Bryan responded: "I believe everything in the Bible should be accepted as it is given there; some of the Bible is given illustratively" (285).[34] Taking this as a positive response, Darrow tried to show the irrationality of such a position by asking whether Bryan believed that Jonah was really swallowed by a whale. Bryan's response (made after some other sparring) shows the incommensurability of the two positions: "Yes, sir. Let me add, one miracle is as easy to believe as another" (285). When forced to defend the Bible literally, Bryan strategically shifted the issue from rationality to belief. For Bryan, the real issue was not whether the biblical event was rational for a scientist, but whether Bryan was willing to believe it.

The contest continued through a long series of other questions: Did Joshua make the sun stand still? How could there be evening and morning before there was a sun? Where did Cain get a wife? How did the snake walk before he was cursed to crawl on his belly? Was the earth really only 6,000 years old? What did Bryan know of other religions, of ancient history, of the origins of language? Bryan responded to all these questions by asserting his belief in the Bible. For Darrow, each response became another proof of Bryan's irrationality and of the incoherence of a belief in biblical inerrancy. In the middle of the cross-examination Darrow said, "You insult every man of science and learning in the world because he does not believe in your fool religion" (288). He ended, "I am examining you on your fool ideas that no intelligent Christian on earth believes" (304). For Darrow, the creation/evolution issue was a contest between intelligence and stupidity, learning and ignorance, reason and belief.

Although Bryan had very limited control of the cross-examination, he made several comments that show his very different conception of its purpose. At one point another prosecuting attorney objected to the questioning, asking: "What is the purpose of this examination?" Bryan responded: "The purpose is to cast ridicule on everybody who believes in the Bible" (299). Darrow strongly objected to this statement of his position, insisting, "We have the purpose of preventing bigots and ignoramuses from controlling the education of the United States and you know it, and that is all" (299). At this point the two attorneys were clearly battling over whose account of their different purposes should prevail. Bryan did not retreat: "I am glad to bring out that statement. . . . I am simply trying to protect the word of God against the greatest atheist or agnostic in the United States. I want the paper to know I am not afraid to get on the stand in front of him and let him do his worst. I want the world to know" (299). A few moments later he put his purpose directly in terms of *belief*: "The reason I am answering . . . is to keep these gentlemen from saying I was afraid to meet them and let them question me, and I want the Christian world to know that any atheist, agnostic, unbeliever, can question me any time as to my belief in God, and I will answer him" (300). From Bryan's perspective, this trial was not a contest between a fact and a belief but a contest between competing beliefs. In the schoolroom of John T. Scopes, the belief in evolution was unfairly attacking the belief in creation, just as in the courtroom, Darrow was now unfairly attacking the belief in God. Bryan was defending the beliefs that seemed most rational to him.

The cross-examination ended when the court was adjourned for the day. The next day, the judge decided to strike Bryan's entire testimony from the record and to forgo any further examinations by either side. After this development, Bryan expressed a desire "to answer the charges made by the counsel for the defense as to my ignorance and my bigotry" (307). Knowing that Darrow had been able to portray him in a negative light before the press, he indicated that he wanted to submit to the press "the questions that I would have asked had I been permitted to call the attorneys on the other side" (301). He explained his request as follows: "I think it is hardly fair of them [the defense] to bring into the limelight my *views on religion* and stand behind a dark lantern that throws light on other people, but conceals themselves. I think it is only fair that the country should know the *religious attitude* of the people who come down here to deprive the people of Tennessee of the right to run their own schools" (308). Bryan did not want to attack

Darrow's rationality but rather his "religious attitude." For Bryan, even the previous cross-examination focused on his own "views on religion" rather than on his powers of rationality. Now he wanted to focus the light of examination on the crucial issue for him—belief—from the other side.

Bryan's questions reveal a significant difference between creationist and evolutionist worldviews as evidenced in this trial.[35] They included the following: "Do you believe in the existence of God as described in the Bible?" "Do you believe that the Bible is the revealed will of God, inspired and trustworthy?" He continued with many other questions that revealed, in their main verb, *believe*, his most important value and the central issue of the trial from his perspective, to which he had strategically shifted from the question of strict rationality. Did Darrow believe in Christ? In miracles? In immortality? His last question was perhaps the most revealing of all: "If you believe in evolution, at what point in man's descent from the brute is he endowed with hope and promise of a life beyond the grave?" Bryan is able to see a final disproof of Darrow's theory of evolution in the fact that men now have a Christian hope of glory.

It may be hard for contemporary readers to conceive of Bryan wanting to ask these questions in a public trial before a thousand witnesses and the journalists of the world. The questions reveal some of the differences in worldviews between the sides in this celebrated confrontation. Evolutionists reading about this cross-examination, and thinking in terms of their own meanings and valences for *reason* and *belief*, would probably find it hard not to see Bryan as an enemy of learning, freedom, and science—but creationists saw him as a Defender of the Faith, a courageous person attacking the dangerous belief in evolution, which was clearly irrational because it contradicted the Bible and left out God. To evolutionists Bryan may have been proving his irrationality by believing in the inerrancy of the Bible, but to creationists he was rationally arguing from the fact that the Bible is true.

Many people besides the creationists have applied the true/false dichotomy to the status of the Bible. Some have resolved the creation/evolution controversy by deciding that the Bible is false. Others have posited alternate dichotomies to correlate with the notion of truth as applied to the Bible. One common solution has been to call the Bible *figurative* rather than *literal*. Darrow himself used this strategy during the trial: "We say that 'God created man out of the dust of the earth' is simply a figure of speech" (188). This solution allows one to retain the term *true*, but only by correlating it with the literal/figurative dichotomy, another hierarchy in which *figurative* has most often been the devalued term. To call the Bible figuratively true is thus not to resolve the problem, but to shift it to a different ground and to give the term *truth* a different meaning. By using this or other dichotomies as rhetorical tools, those who believe in the Bible but reject it literally have attempted to explain what they mean when they say the Bible is true.[36]

This issue of biblical inerrancy has been complicated in the United States in the past century, not just by the true/false and the literal/figurative oppositions, but by the status of the terms in the basic opposition of reason/belief. Since the Enlightenment, the United States and most other Western nations have decided to resolve questions of law and policy not on the basis of beliefs, but on the basis of reasons. As a result, much more has been articulated about processes of reasoning than about processes of believing. Indeed, as the devalued term in this particular dichotomy for at least the last 400 years (if not the 2,000 years since Plato), *belief* has been subject to much less scrutiny.[37] Such scrutiny would probably help people to account for beliefs in less individualistic and more consistent, "rational" terms.

A prime target for such a study would be the belief in biblical inerrancy. (Only a brief, preliminary attempt will be made here.) How would fundamentalists defend their belief in biblical inerrancy as rational? If one means by "rational" something like "consistent," then a possible answer to this question involves three points. First, inconsistencies in the Bible do not present themselves for solution. Indeed, an inconsistency must be perceived before there is anything to be resolved. Because fundamentalists are convinced that such inconsistencies do not exist, they do not attempt to find them. When an occasional inconsistency is found or pointed out, they do not just surrender their belief in biblical inerrancy; instead, they attempt to account for the inconsistency in some other way. James Barr writes that fundamentalists regularly shift between literal and figurative meanings so as to keep the Bible free of errors.[38] This strategy resolves a vast number of apparent inconsistencies.

Besides, fundamentalists are seldom fully consistent in their belief in inerrancy, anyway. Bryan has been chided by later authors for his inconsistency, both in admitting during the Scopes cross-examination that he did not believe in the six literal days of Genesis and in admitting in private, but not in public, that he could accept evolution as long as it did not account for the origin of humanity.[39] Bryan and other creationists draw their lines differently when it comes to the question of which passages in Genesis completely contradict evolution and which ones can be resolved within an evolutionary account. Such differences prove, not that inerrancy is irrational, but rather that people mean different things by *inerrancy*. They interpret it differently in relation to different passages and have their own, differing standards of consistency.

Finally, fundamentalists, and people in general, are probably not as consistent as they would like to think. People regularly hold inconsistent notions without noticing until, for some reason, the inconsistency becomes obvious or problematic. A person can simultaneously believe that the Bible is inerrant and that evolution is true without necessarily having to correlate these two beliefs; only if the beliefs are challenged will people stop to analyze them and attempt to make them cohere. These three points about inerrancy and consistency all suggest that the perception of inconsistency is itself a perception, and not an obvious fact; inconsistency is only perceived against a frame that makes it both noticeable and significant. A person can be consistent in regard to other issues without being consistent in believing in the inerrancy of a particular Bible verse. An inconsistent belief in inerrancy does not prove a person's irrationality in general unless the issue of consistency has been defined as the basis for rationality, as it was in the Scopes Trial cross-examination.

In spite of Bryan's cross-examination, the Scopes Trial ended with a conviction. This verdict was no surprise considering that even Darrow himself had called for it in his summation. The defense expected from the start to lose the battle in the local court because of the religious beliefs of the Dayton residents. They wanted to appeal the case so that the law could be declared unconstitutional, thus warning the rest of the states not to infringe on academic freedom by passing similar laws. They were very disappointed when the conviction was overturned on a technicality and then dropped by the state (which had gotten far more publicity from the trial then it had bargained for).

However, the Scopes battle was far from over; in fact, a lengthier battle remained to be fought. This was the battle over how the Scopes Trial would be represented, what the case would mean; it concerned whose account of the trial and its significance, and told to whom and for what purposes, would exercise the greatest persuasive power in the future. These later versions of the trial use many of the same dichotomies to revisit some recurring conflicts in the West between philosophy and rhetoric. They also repeat a recurring theme of the creation/evolution controversy: that rhetoric poses a threat to civilization.

THREE 1920s ACCOUNTS OF THE TRIAL: DARROW, MENCKEN, AND LIPPMANN

One important set of accounts of the Scopes Trial consists of the newspaper reports sent from Dayton. The day-to-day events of the Scopes Trial were narrated by more newspaper reporters than any previous event in

American history. As one of America's first "pseudo-events" (to borrow a term from Daniel Boorstin), the trial was a deliberate attempt to generate this interest; both sides wanted to use it to influence not just a Dayton judge and jury, but the entire American public.[40] Bryan and the prosecution wanted to persuade the public that evolution is a dangerous belief which should not be taught in public schools, whereas Darrow and the defense wanted to keep the public from passing any more antievolution laws so that evolution could be taught in the schools as scientific knowledge. Not surprisingly, most journalists took Darrow's side on this issue; they portrayed the trial as a contest about Scopes's freedoms of speech and thought, especially after the trial was decided in favor of the state. Their voluminous newspaper reports of the trial will not be treated here. In spite of all these reports, the creationists continued to pass antievolution laws for several more years, and the evolutionists and reporters continued to oppose them.

Once the trial was over and the newspaper reporters had gone home, the persuasive contest shifted to accounts by the participants themselves. How did the participants interpret the trial? What stories did they tell? Of course they did not wait to shape their stories until the trial was over; interpreting the trial was part of the work of the trial itself. Darrow clearly tried to shape the trial by cross-examining Bryan, thus attempting to portray it as a contest about rationality. Bryan did not have a similar chance to cross-examine Darrow and thus could not directly undercut this particular act of persuasion, but he planned to undertake a persuasive labor of his own: to deliver a new speech against evolution as his summation for the jury. Bryan's speech was prevented from reaching millions in the radio and newspaper audiences when the prosecution and the defense both agreed to dispense with closing remarks; even Darrow wanted the jury to find Scopes guilty so they could go on to the appeal and attempt to get the law overturned. This agreement deprived Bryan of the chance to give the oration on which he had been working for months. Although this agreement was mutually beneficial to both sides for the trial itself, it clearly benefited Darrow more than Bryan in terms of later trial representations; it allowed Darrow to make his case in his own best genre of cross-examination and denied Bryan the chance to respond in his best genre, the prepared speech.[41] Indeed, had Bryan delivered this speech, a slightly different story of the trial might have been told overall.[42]

Bryan was very disappointed not to give his speech, but he did not intend to let Darrow's story prevail. Immediately after the trial, he arranged for the speech to be published and planned to travel the country, delivering it to audiences everywhere; but five days later, before he had any other chance to give his version of the events, he died. Many people have speculated that the stress of the trial killed him. Willard Smith reports that "when asked to comment on his view of Bryan at Dayton, [the journalist H. L.] Mencken replied: 'Well, we killed the son-of-a-bitch.'"[43] As a result of Bryan's sudden death, his reaction to the trial is still somewhat of a mystery. How did he feel about the Darrow cross-examination? What was his reaction to the verdict? Of course, he gave some hints at the time, but these are still unanswered questions. In fact, the very uncertainty about Bryan's reactions has become a principal topic for subsequent narrations of the trial. Did he feel defeated or victorious, angry or relieved? Was he more or less dogmatic about evolution? More or less committed to biblical literalism? These answers themselves have become valuable chips in the poker game of whose account of the trial shall prevail.

Darrow, the other key actor, was able to give his version of the Scopes Trial for many more years. As one would expect from his performance in the trial, he hardly portrayed it as a defeat for evolutionists. His best-known account takes up three chapters in his autobiography, The Story of My Life.[44] In these chapters he reveals his own key commitments, to education, intelligence, freedom of speech, and religious tolerance. Assuming the obviousness of the truth of evolution, he depicts Bryan as a demagogue, creationism as an irrational position, and the trial as a battle between correct and incorrect definitions for all the key terms. Darrow sees creationists as people who oppose the truth rather than people who disagree about what counts as truth.

Darrow weaves a powerful linguistic net around the term *truth*. In explaining the background to the trial, he says that Scopes "was indicted for the crime of teaching the truth" (248) by a town that wanted "the honor of prosecuting the boy for teaching science" (254) and under a law that made "the teaching of science a criminal offense" (260). Darrow's first step is thus to equate truth and science. Then he brings in a third synonym by calling this law part of "Mr. Bryan's campaign against knowledge" (248). These three synonyms are then linked to *education* and threatened by another abstraction: "I knew that education was in danger from the source that has always hampered it—religious fanaticism" (249). How did Darrow plan to prove that education provides truth, science, knowledge, and evolution? He reports that his strategy was "to introduce evidence by experts as to the meaning of the word *evolution* and whether it was inconsistent with *religion* under correct definitions of both words" (260). Darrow explains that the truth will become obvious if the defense can only "prove the meaning of the words" (265).

How does Darrow account for Bryan's inability to understand these words correctly? He attacks his rationality: rather than "representing a real case [as the Scopes prosecutor], Bryan represented religion, and in this he was the idol of all Morondom" (249). Darrow adds that, in contrast to the religious beliefs that formed the spurious contents of Bryan's mind, "As to science, his mind was an utter blank" (250). Darrow intended to fill this blank during the Scopes cross-examination; as a result, Bryan "would have been compelled to choose between his crude beliefs and the common intelligence of modern times" (267). However, instead of thinking during this time, Bryan "twisted and dodged and floundered" (267), proving that he would rather be irrational than give up his beliefs. Darrow's rhetoric suggests that beliefs are falsehoods and that only an unintelligent person could fail to tell the difference between an irrational belief and knowledge based on reason.

Darrow similarly attacks Bryan's intelligence in his descriptions of the latter at the trial. At the start of the trial he indicates that Bryan did not look like a hero but "like a commonplace fly-catcher," swatting flies off "his bald, expansive dome and bare, hairy arms" (257). At the end of the trial Darrow no longer saw in Bryan's face "the pleasing smile of his youth," and the "merry twinkle" of his eyes but "his huge jaw pushed forward, stern and cruel and forbidding, immobile and unyielding as an iron vise" (277). As the reason for this change, Darrow asserts: "His speculations had ripened into unchanging convictions. He did not think. He knew. His eyes plainly revealed mental disintegration" (277).

Besides maligning Bryan's intelligence, Darrow uses other strategies of persuasion in his account of the Scopes Trial. In particular, he briefly introduces two image patterns that have been used repeatedly in other representations of the trial, patterns that link Bryan's voice and his appetite to his rejection of evolution, and finally to rhetoric itself. The patterns repeat distinctions made between rhetoric and philosophy many times throughout Western history: they attack the language used to attain power and defend the language used to convey truth, linking the first to creationism and the second to evolution.

Bryan's voice has captured many imaginations besides that of Darrow. Indeed, it was probably the detail about Bryan that was most regularly mentioned by his contemporaries. For example, after hearing Bryan speak, John Reed wrote: "He spoke slowly as if appealing to our reason, not our emotions. And yet the rising and falling of that extraordinary voice, the way he played upon the sentimental prejudices of common men, was what got them most of all."[45] In an age before the microphone was invented (the first newscast of an American trial came from the Scopes Trial itself),[46] Bryan could make himself heard outdoors to a crowd of 30,000 with such clear enunciation that everyone could understand him.[47] He always spoke to capacity crowds. His opponents who went to hear him sometimes switched to his side before the end of his speech because they were so impressed by his extraordinary voice and his great speaking ability.

Darrow heard this voice not only in the Scopes courtroom, but in local churches, where Bryan spoke between trial sessions. Choosing a church as the setting for his description, Darrow represents this extraordinary voice as a lion's roar: "Bryan's voice would boom out above all other sounds and the audience would be as silent and awed as the other denizens of the wilds were when the lion was abroad at night proclaiming himself king of the jungle realm" (262). Darrow compares the power of Bryan's voice over his listeners in the church to the power of the king of the jungle and later suggests that this image has been working in his imagination he compares Bryan speaking in the Scopes courtroom to "a wild animal at bay" who would "commit any cruelty that he believed would help his cause" (277). For Darrow, Bryan's voice is impressive but also potentially dangerous; it inspires awe in the church but may also lead to cruelty in the political arena. Calling up images of other dangerous speakers throughout history who have aroused the masses through their rhetoric, Darrow adds: "History is replete with men of this type, and they have added sorrow and desolation to the world" (277).

Bryan's other memorable personal quality was his hunger. He was legendary for his large appetite and for his willingness to eat before other people; he caused no small stir as secretary of state by bringing a lunch box with him to the State Department every day and by regularly snacking on radishes.[48] Part of his appetite apparently came from his inexhaustible energy; during one presidential campaign, he made up to thirty speeches a day and traveled 18,000 miles in a period of four months. In Michigan alone he traveled 1,000 miles and gave sixty-six speeches in only four days.[49] His energy and his appetite continued throughout his life.

In the autobiography, Darrow mentions Bryan's appetite only once. He writes that on the day of Bryan's death, Bryan ate a large dinner and then lay down for the nap, from which he never awoke: "The too generous meal, with the thermometer around a hundred degrees, brought on his death" (270). This detail is significant because it was not corroborated by Bryan's personal physician[50] nor by his wife, but was nevertheless repeated by Darrow.[51] When told by someone that Bryan had died of a heart attack, Darrow less cautiously responded: "Broken heart nothing. He died of a busted belly."[52] Darrow suggests the significance of this detail when he writes that "a man who for years had fought excessive drinking lay dead from indigestion caused by overeating" (270). For Darrow and others, Bryan's appetite was ironic, perhaps almost hypocritical; it proved that the staunch prohibitionist had physical as well as spiritual desires. The implication was that a man who ate as heartily as Bryan could not be as otherworldly as he claimed. In repeating this *ad hominem* argument, other representations of Bryan make less subtle connections between the man's appetite, his voice, his opposition to evolution, and his desire to attain popular applause and political power through his rhetoric.

These images and themes regarding Bryan's voice and his appetite were amplified by H. L. Mencken, the well-known journalist, writer, and self-proclaimed cynic and atheist. Mencken covered the Scopes Trial as a reporter for all but the last day. He also wrote an antieulogy for Bryan after the unexpected death. In these representations Mencken is much more direct than Darrow in attacking Bryan and depicting the political threat of his creationism through images long associated with rhetoric.

In the trial reports, Mencken suggests the difference between creationists and evolutionists by comparing Bryan to one of Darrow's scientists. He writes that Bryan saw in the scientist "a sworn agent and attorney of the science he hates and fears—a well-fed, well-mannered spokesman of the knowledge he abominates."[53] In contrast to this man of science, Mencken describes Bryan as an enemy of learning, a man of "peculiar imbecilities" (583), a dangerous "fanatic, rid of sense and devoid of conscience" (600). Mencken warns the reader against such a person: "But let no one, laughing at him, underestimate the magic that lies in his black, malignant eye, his frayed but still eloquent voice" (593). Bryan's rhetoric is represented as black magic, which is embodied most powerfully in his eye and his voice.

In reporting on the fifth day of the trial, Mencken gives another description that can be taken as a brief essay on rhetoric. He explicitly calls the two long speeches given by Bryan and Dudley Field Malone "yesterday's great battle of the rhetoricians" (593). Then he writes that Malone defeated Bryan in this battle with a speech that was "simple in structure" and "clear in reasoning," which caused the crowd to give a cheer "at least four times as hearty as that given to Bryan" (594). For a moment Mencken seems to suggest that the plain language long associated with philosophy was able to defend the truth of evolution, but then he explains this positive reaction from the crowd: "For these rustics delight in speechifying, and know when it is good. The devil's logic cannot fetch them, but they are not above taking a voluptuous pleasure in his lascivious phrases" (594). Mencken suggests that Malone's speech was admired not for its logic, but for "lascivious phrases," which give "voluptuous pleasure." In linking language with illicit pleasure, Mencken is repeating a characterization that is common throughout Western history: rhetoric as the bringer of forbidden pleasures, most often as a whore. Darrow figures Bryan's rhetoric as gluttony; Mencken figures it as lust.

Mencken's eulogy for Bryan, "In Memoriam: W.J.B.," uses further images of his voice and appetite to attack the man's creationism.[54] Echoing Darrow's description of Bryan as "a commonplace fly-catcher," Mencken begins the eulogy by asking: "Has it been duly marked by historians that the late William Jennings Bryan's last secular act on this globe of sin was to catch flies?" (64). This sentence uses the words *secular* and *sin* to link Bryan's actions with politics and religion and then undermines these actions by linking them (echoing Darrow) to catching flies. Mencken adds that Bryan was not only "the most sedulous fly-catcher in American history, [but] in many ways the most successful" (64). In this case Bryan's success in catching flies was his ability to advance his political agenda and his religious beliefs by defending creationism against evolution.

Mencken accounts for this success in terms used at least since Aristotle to attack rhetoric: by attacking Bryan's popular audience. He writes that Bryan "liked people who sweated freely, and were not debauched by the refinements of the toilet," whose wives were "as fecund as the shad" (64–65). These people came in large numbers to see the Scopes Trial, where Bryan was surrounded not by educated men and women, but by "gaping primates" who themselves proved the theory of evolution (65).[55] In these vivid images Mencken forges linkages (similar to those drawn by Darrow) between rhetoric and the beasts of the jungle, thus suggesting that rhetoric appeals to our subhuman side.

How was Bryan able to gather these people and solidify their power? He did so, Mencken says, directly through his voice: "He knew every country town in the South and West, and he could crowd the most remote of them to suffocation by simply winding his horn" (64). Bryan's voice becomes a winding horn, a blaring and monotonous sound which brings in vast crowds to hear him speak among whom one cannot breathe. Throughout the rest of the eulogy, Mencken develops this motif of Bryan's voice, which, alternately figured as enchantment and disturbance, is what creates the political threat.

Mencken makes several other attacks on Bryan in terms often used to attack rhetoric, including discussions of his ambition and insincerity. A final attack on Bryan's rhetoric suggests the danger of allowing people like him to have political power. Expressing gratitude that the United States was then governed by the boring Calvin Coolidge rather than by Bryan's "intolerable buffoonery," Mencken writes: "We have escaped something—by a narrow margin. . . . That is, so far" (72–73). He describes the remaining danger: "Heave an egg out of a Pullman window, and you will hit a Fundamentalist almost everywhere in the United States today. . . . They are everywhere where learning is too heavy a burden for mortal minds to carry, even the vague, pathetic learning on tap in little red schoolhouses" (74). Bryan's rhetoric is, finally, not just a source of humor, but a threat; the creationism of fundamentalists like Bryan endangers learning itself.

In this eulogy Mencken thus claims that his own position uses the language of truth, whereas the creationists use the language of political and religious interests. He does not take the next step and call for the prohibition of creationism as the one idea too dangerous to be allowed free expression. (Such a step was taken by the evolutionists in the Arkansas Creation-Science Trial, as described in Chapter 4.) However, the eulogy uses many other vivid rhetorical strategies to portray evolution as an obvious truth and creationism as an irrational belief advancing a dangerous political agenda.

Mencken's and Darrow's accounts of the Scopes Trial have been treated at length because these two writers actually took part. Two other participants also wrote essays about the trial, but they deal only incidentally with Bryan.[56] From the anti-Bryan perspective, many contemporary accounts of the trial were written by nonparticipants and will not be considered here. Those accounts that I have examined trace the same terminology battles and make use of the same correlations between creation/evolution, truth/error, and rhetoric/philosophy.[57]

The one sympathetic treatment of Bryan by a contemporary was journalist Walter Lippmann's book *American Inquisitors: A Commentary on Dayton and Chicago*. Lippmann attempts to ask important questions raised by the trial, mainly by constructing entertaining dialogues between imaginary speakers such as Jefferson, Socrates, and Bryan. Along with Darrow and Mencken, Lippmann sees freedom as the central issue in the Scopes Trial; he writes that the trial "marked a new phase in the ancient conflict between freedom and authority."[58] Lippmann also recognizes the dilemma of forcing people like Bryan to support the teaching of an idea they consider to be dangerous and false. He describes the trial not as a simple fight between truth and error, but as a complex battle between two different systems of beliefs. In contrast to Bryan's creationism, he suggests that Jefferson and the other founders of the United States had what amounts to a faith in the power of human reason, and thus supported a "religion of rationalism" (19). The problem with this "religion of rationalism" (which could also be described as the religion of science) is that it stresses doubt and tolerance without seeing the limits that are implicit in these two notions. Lippmann contends that one cannot be doubtful and at the same time believe in the Bible as God's revelation: "If you are open-minded about revelation you simply do not believe in it" (64). Regarding tolerance, he adds: "Reason and free inquiry can be neutral and tolerant only of those opinions which submit to the test of reason and free inquiry. . . . This disagreement [between reason and revelation] goes to the very premises of thought, to the character of thinking itself. It revolves upon the question of whether human reason is or is not the ultimate test of truth for men" (86). Lippmann suggests that rationalism necessarily limits its tolerance to those positions that recognize its superior claims. Therefore, those positions that do not submit to reason, such as creationism, are labeled intolerant by definition and thus are excluded from the dialogue from the start. Thus, even reason itself has a particular perspective of its own, which creates its own blind spot; it labels irrational and dangerous those positions that will not submit to its tests. It cannot tolerate intolerant positions.

This insight provides a key to understanding the conflicting worldviews of creationists and evolutionists. The two groups put their faith in different first principles: the creationists, in a revelation of God through a literal Bible, and the evolutionists, in reason itself as a method of learning the truth in nature. Both groups reject the opposite position as a betrayal of their own faith. Their battle is not, finally, a battle about details, but one about frameworks within which to put details. In a culture constructed on Enlightenment models of rationality, it is exceedingly difficult to know what to do about claims for revelations from God, even when those claims are made in the name of reason itself.

Lippmann ultimately shares the commitments of Mencken and Darrow; he believes that political freedom in the United States can be defended only on the basis of reason as these men conceive of it. In the framework of his own commitment to liberal institutions, he concludes that the Scopes Trial did indeed pit reason against revelation and that Bryan was an inquisitor after all.[59] Except for Lippmann's sympathetic treatment of the creationist problem of revelation, Bryan's other contemporaries continued to describe the story as a battle between truth and error, in which rational, tolerant heroes fought against irrational, intolerant villains who were led by rhetoric to accept mere beliefs.

A 1950s ACCOUNT OF THE TRIAL:
INHERIT THE WIND

The common 1920s view of the Scopes Trial as a battle between good and evil seems to have been the dominant story told about the trial in the

United States until the end of World War II.[60] By pitting obvious truth against obvious error in this way, the story implied that knowledge would gradually overcome ignorance, and thus, that creationism would gradually fade away as the truth of evolution became more clearly established. The story also tended to depict the trial as a decisive defeat for creationists and the beginning of a decline for American fundamentalism.

However, these implications were not borne out in any detail. Religious historians such as Ferenc Morton Szasz point out that in fact, the antievolution campaign was even more active after Bryan's death. Fundamentalism clearly did not disappear, but only regrouped.[61] Several studies have also shown that high school biology textbooks significantly lessened their emphasis on evolution after the trial because of the strength of public opinion against it.[62] Edward Larson concludes: "The bewildering disparity in opinion about the impact of Scopes on the anti-evolution movement . . . highlights the unexpected resilience of this cause despite its pounding at Dayton."[63]

The Scopes Trial began to seem important again in the early 1950s, after the onset of McCarthyism and the Cold War. As a result of increasing political upheaval and suspicion during this time, a new outpouring of scholarly work attempted to make sense of the 1925 trial. What to many had seemed an aberration in American history now began to look like the repetition of a pattern; many scholars wanted to explain what they took as a new outbreak of irrationality by understanding the outbreak of thirty years ago. The most influential representation of the Scopes Trial produced during this time period was the play *Inherit the Wind* by Jerome Lawrence and Robert E. Lee. The publishing history of the play was directly influenced by its political milieu.

After twelve years of intermittent research and writing, Lawrence and Lee finished drafting the play in 1950, but they decided not to publish it at that time because, as Lee told a reporter who interviewed them in 1955, "the intellectual climate was not right."[64] In February 1950 Senator Joseph McCarthy had begun his anticommunist campaign with a famous speech denouncing 205 communists he claimed were working for the State Department. "At that time," Lee later reported, "I wouldn't have dared write a letter to my Congressman."[65] For four years Senator McCarthy exercised considerable power as the chair of a powerful Senate Investigations subcommittee, which held extensive hearings on suspected communists, but he was finally discredited for attacking U.S. Army officials without proof and was censured by the full Senate in December 1954. Only one month after McCarthy's demise, *Inherit the Wind* was staged, first for three weeks in

Dallas and then at the National Theater in New York, where it was a hit for two years. It was made into an award-winning movie in 1960.

In their preface to the play, Lawrence and Lee suggest that these political conditions influenced their treatment of the Scopes Trial. They write: "*Inherit the Wind* is not history. The events that took place in Dayton, Tennessee, during the scorching July of 1925 are clearly the genesis of this play. It has, however, an exodus entirely its own."[66] After thus disclaiming that they are giving a factual account of the trial (and joking already about Genesis versus Exodus), they add that they have changed the setting and the names of the characters. They suggest one purpose for their play in an ominous conclusion: "[The play] does not pretend to be journalism. It is theatre. It is not 1925. The stage directions set the time as 'Not too long ago.' It might have been yesterday. It could be tomorrow" (ix). Rather than returning the play to an incident of thirty years ago, the playwrights make it contemporary: it could happen yesterday or tomorrow. They want to use the trial primarily to present their own themes and relate them to their own milieu.

The play can be analyzed as an extended argument for free thought in the 1950s—for people to use reason in order to question the irrational beliefs and rhetorical tactics of people like McCarthy—but it makes this argument through a Scopes-like trial opposing creationism and evolution. As a result of this focus, the work deserves a close reading for its treatment of the specific argument between creation and evolution, its representations of Bryan as rhetorician, and its analysis of the roles of reason and belief in the United States.[67]

The work begins by setting up a major conflict between reason and belief as embodied in creationism and evolution. At the start, Bertram Cates, the character loosely based on John T. Scopes, has been arrested for teaching evolution and placed in the city jail. In the movie his arrest is especially dramatic; three political officials and the town minister invade Cates's classroom just as he begins to teach his high school students from Darwin's *Origin of Species* (while standing in front of a chart of a male gorilla) that we "evolved from a lower order of animals, from the first wiggly protozoa in the sea to the ape, and finally to man." Scopes's arrest is photographed and becomes the front-page story for many national newspapers. In this work, the arrest symbolizes a widely watched legal confrontation between a man threatened for teaching the truth of science and the forces of politics and religion arrayed against him.

The major thematic vehicle of the work is a love plot: Bert Cates is engaged to Rachel Brown, the only daughter of the town minister who helped arrest him. The plot follows the unfolding relationship between Bert and

Rachel, tracing, through their life together and their relationships to other key characters, the argument between creationism and evolution. Near the start, Rachel visits Bert clandestinely in the jail, bringing him clean clothes and begging him to apologize for teaching evolution before it is too late. Bert reassures her: "All [Darwin's *Origin of Species*] says is that man wasn't just stuck here like a geranium in a flower pot; that living comes from a *long* miracle, it didn't just happen in seven days" (7, italics in the original). After arousing sympathy for this couple, the scene suggests that evolution does not really threaten the idea of creation; it only suggests a different means.

However, others cannot see this easy reconciliation. Rachel's father sees Bert as an infidel and demands in an early scene that she break her engagement to him; Bert later gives her an ultimatum to choose between them: "It's his church or our house." Through two more complications in the plot (one in which her father threatens to curse them both, during a hellfire-and-brimstone, fundamentalist revival, unless they deny the heresy of evolution, and the other in which the Bryan character, who is a sort of temporary foster father for Rachel, betrays her by making her testify against Bert in the trial), Rachel is eventually forced to choose between Bert and her father, and by analogy, between the scientific reasonableness of evolution and her fundamentalist faith in biblical creation. She chooses Bert and evolution, and at the end, they leave the small town of Hillsboro to set out on a life of their own. Through these main characters and events, the work argues that creationism must eventually give way to evolution.

The play also spatially represents the conflict between creationism and evolution in the image of the houses between which Rachel must choose. The title of the play comes from a verse in Proverbs: "He that troubleth his own house . . . shall inherit the wind" (60). That which troubles Rachel and makes her choose between the house of her father and Bert's house is creationism itself, depicted as a strict and inhumane belief that causes her own father (as well as her foster father, the other key creationist) to lose his daughter, and thus to inherit the wind. The house is an outstanding symbol of this conflict because of its biblical associations with phrases such as "the house of Israel"; indeed, even in its title, this work uses biblical imagery and phrases extensively to persuade its audience not to believe the Bible literally. This particular symbol suggests that the house of the fundamentalists—their faith, tradition, lineage, and posterity—is threatened unless they abandon their strict biblical faith and accept the truth of evolution.

However, the townspeople resist this truth and anyone who would teach them about it. They violently threaten Bert and his defense attorney—the Clarence Darrow character, who is named Henry Drummond (played in the movie as a kindly and sympathetic man by Spencer Tracy)—demanding that these men stop teaching this idea, which seems to them to oppose God. Wearing banners and toting placards against evolution, the townspeople in various scenes tell Bert, "We'll ride you out [of town] on a rail"; gather in a group to sing, "We'll hang Bert Cates to a sour apple tree"; throw a bottle at him, which breaks on the wall of the jail and cuts his face; and burn his effigy (and Drummond's) in a scene intended to remind a 1950s audiences of the Ku Klux Klan (in the very next scene, the H. L. Mencken character, journalist E. K. Hornbeck, comes to Drummond's room wearing a white hood).[68] Using these powerful images, the movie represents the townspeople as a lynch mob ready to kill, with the implication that the creationists are willing to defend their belief and oppose the truth of evolution to the death.[69]

The work makes the same point in its treatment of the William Jennings Bryan character, Matthew Harrison Brady (who is played in the movie as a strutting and distasteful man by Fredric March). Brady is an outspoken enemy of science who says he wants "to test the steel of our Truth against the blasphemies of Science" (21) and insists that "the way of scientism is the way of darkness." He is a biblical literalist and absolutist who asserts, during his cross-examination by Drummond, that he will never read Darwin, that he trusts the Bible more than the solar system (79), and that he knows when the creation actually occurred (on October 23, 4004 B.C., at 9:00 A.M.) (85). He betrays Rachel by forcing her to divulge on the witness stand damaging confidences that turn the jury against Bert. He seeks vengeance by protesting when Bert is convicted but only fined $100 by the judge; he wants him to be imprisoned as an example to all evolutionists (103). The danger posed by people like Brady is powerfully suggested when Drummond gets Brady to admit on the witness stand that God speaks to him directly and instructs him to pass these revelations on to the world. Drummond says: "The Gospel according to Brady. God speaks to Brady, and Brady tells the world. Brady, Brady, Brady, Almighty!" (89) All these details in the work depict Brady as a political and religious fanatic who considers himself a spokesman for God, and who consequently will defend his beliefs even when they oppose the advancement of knowledge and harm other people.

Through these negative representations of Brady and the townspeople, *Inherit the Wind* suggests that the Scopes Trial pitted the truth of evolution against an irrational and dangerous belief in creationism. In defending the distinctions it draws between truth based on knowledge and erroneous belief, the work turns its interest to language itself, linking Brady's creationism to his voice, his appetite, and ultimately, to rhetoric.

Brady's manner of speaking and his voice are treated in many scenes in the work. When he first arrives in the town of Hillsboro, he gives a rousing speech and then gets his picture taken standing exactly in the middle between the reverend and the mayor. Throughout the rest of the work, his frequent speaking (which is often interrupted or mocked) is regularly linked to politics and religion and distanced from speech involving the truth. At one point in the trial, Drummond says, "I ask the court to remind the learned counsel [Brady] that this is not a Chautauqua tent. He is supposed to be submitting evidence to a jury" (62). The implication is that Brady should save the rhetoric for religious occasions and limit his language now to a presentation of the facts. In a similar scene at the end of the trial, Brady shakes a sheaf of papers and says he has prepared a few brief remarks. Drummond objects: "Mr. Brady is free to read any remarks, long, short, or otherwise, in a Chautauqua tent or a political circuit, but our business in Hillsboro is completed." Rhetoric is welcome in political and religions settings, but it does not belong in a court of law whose language business is done.[70]

In the same scene, Brady attempts to give his speech after the trial has adjourned, but his words are swallowed up in a chorus of voices and other noises, including a boy repeatedly pounding the judge's gavel, a hawker selling ice cream, and a radio announcer returning the live audience to a "matinee musicale" (104–106). Brady frantically tries to shout down the opposition but cannot. Still, he pushes forward to the end, exhorting, "Faith of our fathers, holy faith, we will be true to thee 'till death!" After ending with a call for faith to the death, he himself falls dead, thus ironically suggesting that, in thus coming to an end, his own faith is as hollow as his rhetoric. Especially interesting in this scene is the juxtaposition of a speech supporting creationism with linguistic acts of consumer exchange and entertainment. Whereas philosophy, the "love of wisdom," is regularly portrayed in Western history as imparting the truth without concern for money, rhetoric is regularly portrayed either as the exchange of language for money or as language intended for mere show. In Brady's death scene, talk of faith in creation, of ice cream, and of matinee musicales are all absorbed into a din of meaningless voices now that language has completed its function as an instrument of justice and truth.

The work similarly figures Brady's appetite as a symbol of rhetoric. In an early scene in the movie, Brady is warned by his wife, "Matt, you know what the doctor said about not overeating in the heat." He responds, "I'm gathering strength for the battle ahead." Throughout the movie he continues to gather his strength, eating enormously in many scenes, including one with a group of reporters, and one in the courtroom just before the verdict is announced. The implications of his eating are suggested in this stage direction from the play: "Brady sits eating a lunch. He is drowning his troubles with food, as an alcoholic escapes from reality with a straight shot" (95). Given that William Jennings Bryan was a staunch prohibitionist, this comparison of Brady's appetite to alcoholism almost accuses him of hypocrisy, just as rhetoric has regularly been linked to insincerity. Brady's eating, like his belief in creationism, proves his inability to cope with reality and his desire to use language to escape from the truth.

At least four more scenes in the work can be taken as comments on rhetoric. In the first, Drummond and Brady are talking one night about belief in God while sitting on the front porch. When Drummond says that believers are "window-shopping for heaven," Brady responds that they are only looking for their "golden chalice of hope."[71] These images linking belief, money, and hope remind Drummond of a rocking horse named Golden Dancer that he had wanted as a boy, but that broke when he rode it for the first time because it was "all shine and no substance." Like the shoddy rocking horse, the fine language of rhetoric has been described throughout Western history as shine rather than substance, form rather than content. The scene suggests that religious belief is an unsatisfying illusion, just as rhetoric is a beautiful but hollow package with nothing of value inside, but which is nevertheless for sale in a market economy.

In another scene, Drummond uses a comparison famous in the history of rhetoric. At one point in the trial he tells the jury: "[The theory of evolution] is as incontrovertible as geometry in every enlightened community of minds" (74). One of Brady's fellow prosecutors responds: "In this community, Colonel Drummond—and in this sovereign state—exactly the opposite is the case" (74). Drummond later insists, "This community is an insult to the world." The relevant passage is Aristotle's *Rhetoric*, Book 3, in which he explains that rhetoric only applies to a popular audience (such as this small-town audience that insults the world) and to certain subjects; after all, he writes, "no one uses fine language when teaching geometry" (1404a). Aristotle and Drummond both think that rhetoric is used only to defend beliefs such as creationism; knowledge of the truth is self-evident in the case of geometry and evolution.

A third scene attacking Brady's rhetoric takes place after Brady has collapsed while giving his last speech. As he is carried out of the courtroom, he incoherently mumbles a presidential acceptance speech. Drummond asks: "I wonder how it feels to be Almost-President three times—with a skull full of undelivered inauguration speeches" (108). This quotation suggests that Brady's skull was not full of thoughts but only full of language; the oration that Brady really wanted to give was the speech welcoming him to political power. Drummond implies that the goal of Brady's rhetoric was not truth, but power; all his talk about creationism was not a sincere defense of his beliefs but rather an effort to become president.

The fourth scene summarizes the range of implied attacks on rhetoric made in this work. In this scene the newspaper reporter, E. K. Hornbeck—a caustic character played by Gene Kelly and based on H. L. Mencken—is describing the hordes of people who have come out of the Tennessee backwoods to hear Brady defend his creationism. Hornbeck says: "[They have come] to listen to their plump messiah coo and bellow. Their high priest of mumbo-jumbo, Matthew Harrison Brady, has alternately been stuffing himself with fried chicken and belching platitudes since his arrival here two days ago." This brief passage gives five metaphors for rhetoric: cooing, bellowing, mumbo-jumbo, platitudes, and belches. Brady coos like a pigeon and bellows like a bull to please his subhuman audience. He speaks to them with religious mumbo-jumbo rather than simple and clear words of scientific fact. The fried chicken he eats leads to his belches, just as his appetite for applause leads to his false words. In an earlier scene linking rhetoric and food, Brady also belches, this time while reading a sentence he has written about the Bible to a group of reporters. These images suggest an interpretation of the work's title as a statement on rhetoric itself: Brady is, finally, a creationist/rhetorician rather than an evolutionist/philosopher; he is a bag of hot air, a man who produces only wind. In accordance with its title, the work suggests that anyone who trusts rhetoric or allows it to seize power will inherit the wind.

In attacking creationism as rhetoric and, by extension, as a belief opposed to the knowledge obtained through reason, *Inherit the Wind* could easily be taken as a rejection of belief, and perhaps of religion itself as a legitimate force in American culture. To counteract this implication, the work (especially the movie version) contains other scenes in which it attempts to reconceive, rather than to reject, the term *religion* by using religious language to argue for values that are more crucial to the authors than biblical literalism. Their two most important values are reason and thought—values that are crucial to a liberal democracy founded on the principles of Enlightenment rationalism.

The movie version first hints at this shift in the term *religion* when Brady starts to examine Rachel on the witness stand. When asked whether she and Bert attend the same church, she admits that Bert has stopped attending. Brady asks: "Did Mr. Cates leave the church?" "No," she replies, "Not really. Not the spirit of it." Brady asks, "But the body of it?" to which he eventually receives a subdued "Yes." In this scene Rachel attempts to distinguish between two senses of *church*: the more important, spiritual church and the less important, physical one. The rest of the movie argues that Bert and the other evolutionists are not irreligious, but actually more religious than the fundamentalist churchgoers because they recognize religion as a personal rather than an institutional force.

This recovery of religion is continued in the Brady cross-examination scene. At one point Drummond uses the word *holy*, and Brady asks, "Is something holy to the learned agnostic?" Drummond responds with a climactic speech in which he says that the thing most holy to him is "the individual human mind" (83). Brady then insists that not thought but "faith is the most important thing." Drummond asks, "Then why did God endow us with the power to think?" and praises "our God-given gift of reason." This scene attempts to distinguish between a God who forbids thought and one who encourages it, implying that the latter is a superior conception and that those who both think and believe are the most religious. In such scenes the agnostic Drummond uses the language of holiness and God to argue for the value of human thought and reason.

All the young people in the play also express their struggle with evolution in the valorized language of *thought*. A high school student who testifies against Bert says that he does not know if he believes evolution but will think about it. Bert refuses to apologize for teaching evolution when urged by Rachel, saying he will not tell the officials that "If they [would] let my body out of jail, I'd lock up my mind." Near the end of the work, Rachel confesses to Drummond that she has never really thought before (with the implication that her religious beliefs would not let her). She then gives a simile: "A thought is like a child inside your body. It has to be born. If it dies inside you, part of you dies, too" (111). These passages all privilege thinking over believing by transferring the values associated with life and freedom to the power of thought itself.[72]

In other scenes the work deals directly with the key term *belief*. As with the term *religion*, it does not reject *belief* but only reconceives it. For example, after Rachel has been betrayed by Brady, she comes to his hotel room to confront him. On finding him asleep, she talks with his wife, who tells her that she has naively moved from blind faith in Brady to complete rejection of him, and thus from a position of total belief to one of total disbelief. Brady's wife concludes: "I believe in my husband. What do you believe in?" The scene suggests that people need to believe in something and that a person without belief is pitiable rather than admirable. The work does not want to persuade people to disbelieve everything: it wants them to believe in something other than biblical creation.

The same point is made explicitly in several scenes that contrast Drummond's agnostic uncertainty to Hornbeck's atheistic cynicism. In the final scene of the movie, Hornbeck mocks the dead Brady but Drummond responds, "Matt Brady got lost because he looked for God too high up and too far away." Hornbeck recognizes the religious implications of this statement and accuses Drummond: "You hypocrite. You fraud. The atheist who believes in God. You're just as religious as he was." Drummond does not deny this charge, but instead berates Hornbeck, asking, "Don't you understand the meaning of what happened here today?" Hornbeck responds, "What happened here today has no meaning." At this point in the work, when meaning itself is most radically questioned, Drummond disagrees: "*You* have no meaning. You're like a ghost pointing an empty sleeve and smirking at everything people feel or want or struggle for. . . . What do you dream about? What do you need? You don't need anything, do you? People? Love? An idea just to cling to? You poor slob. You're all alone." A person without any beliefs is all alone in the world, left to perceive the world and even himself as empty signifiers without meaning. Hornbeck tries to recover from this denunciation by praising Drummond for defending his "right to be lonely," but the movie clearly does not admire Hornbeck even though it argues for tolerance regarding people like him. The heroes are Drummond and Cates, the religious men without any church. They both have a cause to which to devote themselves, something to believe in, a source of meaning in their lives that has become what could be called their private religion.

The movie uses one more strategy to redefine *religion*: it starts and ends with different religious songs. The opening song is a rendition by a single alto voice of "Give Me That Old-Time Religion," which is repeated later many times by groups of threatening creationists.[73] The movie ends with "The Battle Hymn of the Republic," sung by the same alto voice, as Drummond picks up his briefcase and leaves the courthouse, carrying copies of both the Bible and the *Origin of Species* as a concrete example of his tolerance.[74] By switching to this hymn, the movie attempts to alter its depiction of religion itself; after attacking religion, as expressed in the first song, it backs away from this attack, portraying Bert and Drummond at the end as religious men who are themselves trying to make sure that the "truth is marching on."

Such negotiations of the key terms *religion* and *belief* suggest that *Inherit the Wind* does not want to reject religion; it just wants to replace certain, intolerant religion, with a tentative, tolerant faith. Drummond says that this change only requires creationists to surrender their "faith in the pleasant poetry of Genesis" (83). However, the creationists disagree with his assessment of the cost; they prefer to see Genesis as literally true rather than pleasantly poetic. Drummond does not want to abandon such terms as *belief*; rather, he wants to shift this term so that it means belief in humanity, tolerance, and reason rather than in the literal Bible. *Inherit the Wind* does not, finally, reject *belief* in favor of *reason*; it reconceives the former so that *belief* and *reason* mesh. The work uses a religious rhetoric to argue for the belief that human reason can discover truth and that American society thus ought to remain committed to a rational, rather than a religious, political ideal. As with other representations of the Scopes Trial, *Inherit the Wind* ultimately uses the terms *truth* and *belief* to defend evolution as a scientific idea and liberal democracy as a political philosophy.

A 1980s ACCOUNT OF THE TRIAL: STEPHEN JAY GOULD

The last example to be considered in this chapter comes from the next great decade of interest in the Scopes Trial, the 1980s. As a result of Ronald Reagan's election to the presidency and the concurrent empowerment of religious fundamentalists such as the Reverend Jerry Falwell and his Moral Majority, contemporary creationists made another effort to defend creationism as a scientific fact and to attack evolution as a belief that unfairly undermined their beliefs (Chapter 4 focuses on one major episode from this decade). This effort created horror among evolutionists, as witnessed in an exemplary retelling of the Scopes Trial by Stephen Jay Gould, a leading evolutionist and a Harvard professor of geology and paleontology. In his essay, "A Visit to Dayton," from his collection, *Hen's Teeth and Horse's Toes* (1983), Gould represents the trial once again as a battle between truth and error that turned on the crucial terms *freedom* and *tolerance*. This final retelling of the Scopes Trial shows some of Gould's key values and commitments. It also illustrates that the details and the tellers change but the story of the abuses of rhetoric remains the same.

As Gould's title suggests, his essay reflects on a visit he made to Dayton in June 1981, as a result of what he calls "our current creationist resurgence."[75] How should one deal with this resurgence? Gould attempts to discredit the creationists by defining all the key terms correctly. He starts by defining *evolution*, contrasting Darrow's mechanical conception, as expressed by the lawyer during the famous Leopold-Loeb Trial, to his own conception of evolution as a random process that "gives meaning to the old concept of human free will" (263). Calling the Scopes Trial a similar "outcome of accumulated improbabilities" (264), Gould prepares to retell the story of this confrontation between the scientific process of improbabilities known as evolution and the improbability that anyone would disbelieve it or find it dangerous enough to oppose its teaching in public schools.

Gould next describes Dayton, relying on many of the details provided in Mencken's news reports while making an implicit attack on another key term, *certainty*. Dayton is still a charming and beautiful city, but "the older certainties may have eroded somewhat" (269). It has what some consider its contemporary problems (for example, a recent marijuana seizure and the availability of condom machines), but at least one old certainty is gone: the belief that evolution causes social evils. Gould writes: "At least they can't blame evolution for this [the condoms], as one evangelical minister did a few months back when he cited Darwin as a primary supporter of the four *p*'s: prostitution, perversion, pornography, and permissiveness" (269–270).[76] Then again, however, maybe the old certainties are not gone, Gould fears. The citizens of Dayton still hold to one such certainty: "They taught creationism in Dayton before John Scopes arrived, and they teach it today" (270).

Having ridiculed two old certainties, Gould turns to the Scopes Trial itself: "The Scopes Trial is surrounded by misconceptions, and their exposure provides as good a way as any for recounting the basic story. In the heroic version, John Scopes was persecuted, Darrow rose to Scopes's defense and smote the antediluvian Bryan, and the antievolution movement then dwindled or ground to at least a temporary halt. All three parts of this story are false" (270). Gould retells the story to reveal these three falsehoods. His strategy as he narrates is to implicitly contrast the term *truth* with those misuses of language that have been grouped at various times in history under *rhetoric*.

Gould introduces the term *truth* directly when he reports that "the potential truth of evolution was not an issue" in the trial (271). He writes that in fact, all the experts who could have settled this question of truth were excluded outright, although they were still able to produce "formidable documents" in support of evolution that "were printed in newspapers throughout the country" and "finally [admitted] into the printed record of the trial" (271), with the implication that the truth of evolution had triumphed. In contrast to Gould's view, Bryan and other creationists thought that evolution was only a belief held by a few experts who were trying to coerce its acceptance without proof. Two of the great questions of the trial were what counts as truth and who qualifies as an expert.

Next Gould turns to Bryan, whose rhetoric he implicitly contrasts with the truth known by the experts. Gould writes that after sitting in "uncharacteristic silence for several days" of the trial, Bryan finally rose to make a "grandiloquent speech" to the rural folk for whom "no art commanded more respect than speechifying." Later, however, "Bryan was just plain outspoken, outgestured, and outshouted" by Malone (271). Then came the cross-examination: "Bryan viewed the occasion as a desperate attempt to recover from Malone's drubbing, but Darrow exposed him as a pompous fool" (272), especially when he admitted that the days of Genesis were not literal, apparently unaware that "local fundamentalists would regard [his admission] as a betrayal, and the surrounding world as a fatal inconsistency" (273). This account of Bryan's performance at the trial depends on the view of rhetoric discussed here in relation to Darrow, Mencken, and *Inherit the Wind*. Gould is convinced that Bryan could not be either rational or serious in opposing evolution. All he was doing was shouting and speechifying, trying to cover up what Gould calls a "fatal inconsistency" (a belief in biblical inerrancy) by using empty language. After indicating that "Bryan was vanquished and embarrassed during the trial" (271), Gould writes: "Bryan recouped [from his humiliation] by involuntarily taking the only option left for an immediate restoration of prestige: he died in Dayton a week after the trial ended" (273). Gould accuses Bryan of acting rhetorically even in dying.

Gould's essay to this point thus narrates the Scopes Trial as a familiar story: in the trial, Bryan used empty rhetoric to defend an irrational belief in biblical creation, but Darrow attempted to use plain language to present evidence for the truth of evolution. Unfortunately, Darrow was thwarted by fundamentalists who preferred their own religious beliefs to scientific truth. Turning the rest of his essay from the Scopes Trial back to the "current creationist resurgence," Gould implies that rhetoric has once again threatened civilization in the guise of creationism and attempts to expose the rhetoric of the recent creationists by separating the scientific truth of evolution from their deceptions of politics and religion.

Gould predicts his thesis for this half of the essay by introducing Kirtley Mather, a Harvard Professor Emeritus and one of the Scopes defense witnesses, whom Gould often invited to his own classes to speak about the Scopes Trial. After establishing Mather's ethos by calling him a "pillar of the Baptist church, lonely defender of academic freedom during the worst days of McCarthyism, and perhaps the finest man I have ever known," Gould writes that Mather's taped lecture has changed from "a charming evocation" to "a vital statement on pressing realities" and that it is time to "dust off the videotape and show it to [his] class as a disquisition on immediate dangers" (273–274). Gould depicts the 1980s creationist effort as an immediate danger and a threat to academic freedom, thus arguing in defense of a value he holds dear.

In specifying the danger, Gould writes that the theory of evolution itself is much more widely accepted than it was during the Scopes Trial. Nonetheless, the creationists are still using devious linguistic strategies to advance their goals: "Evolution is now too strong to exclude entirely, and current proposals for legislation mandate 'equal time' for evolution and for old-time religion masquerading under the self-contradictory title of 'scientific creationism'" (274). Gould suggests that now the creationists are trying to include creation rather than exclude evolution from the classroom, but even though their goal has changed, their methods have not: they are still using faulty definitions and nonsense words.[77] "Creationists are not battling for religion. They debase religion even more than they misconstrue science." He explains their reasons for misusing these two key words: "Creationism is a mere stalking horse or subsidiary issue in a political program that would ban abortion, erase the political and social gains of women by reducing the vital concept of the family to an outmoded paternalism, and reinstitute all the jingoism and distrust of learning that prepares a nation for demagoguery" (275). Gould sees the creationist position, not as an alternate conception of the key terms, but as a willful effort to misuse these terms in support of political goals he opposes. This last quotation suggests very well some of his own key values and beliefs.

Gould finally reveals the central issue of this controversy for an evolutionist committed to liberal democracy: "The enemy is not fundamentalism; it is intolerance. In this case, the intolerance is perverse since it masquerades under the 'liberal' rhetoric of 'equal time.' . . . For all their talk about weighing both sides (a mere question of political expediency), they would also substitute Biblical authority for free scientific inquiry as a source of empirical knowledge" (276). In this passage Gould indicates that for him the crucial issue is intolerance; he holds that creationist intolerance endangers free inquiry. His last phrase suggests that only *scientific* inquiry and *empirical* knowledge are endangered, but the passage does not clarify what other types of inquiry and knowledge he considers valid. If he were to clarify these other types, he might find that he disagrees with many creationists less than he thinks. Indeed, he asserts that he feels threatened by only some of them: "We have nothing to fear from the vast majority of fundamentalists. . . . Rather, we must combat the few yahoos who exploit the fruits of poor education for ready cash and larger political ends" (277). However, the "yahoos" who reject evolution because they want power and money are very dangerous indeed. Gould explains: "Do movements of intolerance ever start in any other way, given our pervasive tendencies toward geniality? Do they not always begin in comedy and end, when successful, in carnage?" (278) What was before merely harmless, ridiculous, irrational (a belief in creationism) becomes a threat of destruction. In the next sentence Gould completes the linkage to Adolf Hitler. He then concludes his essay by warning his readers to heed this warning or face the threat of extermination by the totalitarian creationists.

In this representation of the Scopes Trial, Gould's argument against creationism depends on taking the creationists' beliefs as objects of humor or fear. Gould implies that as long as the creationists are tolerant and their beliefs are amusing, they ought to be tolerated themselves, but if they themselves take their beliefs too seriously and become intolerant, they must be stopped. Within his own framework of values and beliefs, the price may be too high to tolerate the creationists. His conclusion suggests that there is no permanent distinction between knowledge and politics, between his own position and the position of the creationists. Indeed, his own position is political in that it sees creationism as dangerous and asserts that it ought to be opposed. Restating a problem that continually confronts liberal democracies because of this blind spot in tolerance as an ideal, Gould implies that even the word *tolerance* has its limits when one confronts a Hitler; he himself shifts the word, using it as a rhetorical tool. He wants tolerance itself moderated so as to help rather than hurt humankind.

In contrast to Gould's positivist account and the other accounts analyzed in this chapter, a rhetorical account of the Scopes Trial suggests that creationists like Bryan are not people who argue for irrationality and intolerance against learning and freedom. Rather, for these individuals, such important words have different meanings and valences than they do for many evolutionists, as a result of differing key values and beliefs. Both the creationists and the evolutionists inevitably use rhetoric to defend their own meanings and valences for the key terms of this controversy and to reject the meanings and valences of their opponents. Curiously, moreover, they agree on at least one point: they both hate rhetoric, even though they use it in every episode to defend their own position as true and attack the opposing position as erroneous. This is not, finally, a battle between truth and error, but a battle about what counts as truth and error. It is not a battle between rhetoric and philosophy, but a battle between competing rhetorics reflecting competing worldviews.

NOTES

1. Morris Bernard Kaplan, "The Trial of John T. Scopes," in *Six Trials*, ed. Robert S. Brumbaugh (New York: Thomas Y. Crowell, 1969), 107.

2. Ibid., 117. Italics his.

3. Ibid., 112.

4. LeRoy Ashby, *William Jennings Bryan: Champion of Democracy* (Boston: Twayne, 1987), xiv.

5. Quoted in Lawrence W. Levine, *Defender of the Faith: William Jennings Bryan: The Last Decade, 1915–1925* (New York: Oxford University Press, 1965), 154. In the phrase "unrepealable law of the land," Bryan refers to the four constitutional amendments that were ratified with his support.

6. Ibid., 55, 99.

7. Ashby, *William Jennings Bryan*, 67, 92.

8. On the vote, see ibid., 67.

9. Quoted in ibid., xvii.

10. Quoted in Edward J. Larson, *Trial and Error: The American Controversy over Creation and Evolution*, updated ed. (New York: Oxford University Press, 1989), 31.

11. William Jennings Bryan, "The Prince of Peace," in vol. 2 of *Speeches of William Jennings Bryan* (New York: Funk and Wagnalls, 1913), 261–290. Subsequent page references to this speech appear in parentheses in the text.

12. An account of Bryan's developing opposition can be found in Ferenc Morton Szasz, *The Divided Mind of Protestant America, 1880–1930* (University: University of Alabama Press, 1982), 110–111.

13. William Jennings Bryan, "The Origin of Man," in *Seven Questions in Dispute* (New York: Fleming H. Revell, 1924), 123–158. Further page references to this speech appear in parentheses in the text.

Another important speech is William Jennings Bryan, "The Menace of Darwinism," in *In His Image* (New York: Fleming H. Revell, 1922), 86–135. This book is a collection of nine speeches originally presented by Bryan at the Union Theological Seminary in Richmond as the James Sprunt Lectures. "The Menace of Darwinism" appears as Chapter 4, "The Origin of Man." It was later renamed, reprinted separately, and delivered hundreds of times under the new title.

A third antievolution speech was "The Bible and Its Enemies" (1921), which was originally delivered to the Moody Bible Institute of Chicago and published under the same title in Chicago by the Bible Institute Colportage Association. This last speech lists four enemies of the Bible: the Agnostic, the Atheist, the Higher Critic, and the Evolutionist. Of the four, Bryan called the Evolutionist the worst (19). Both "The Menace of Darwinism" and "The Bible and Its Enemies" contain the same major argument against evolution as "The Origin of Man," but they both give expanded examples delivered in a more rancorous tone.

14. In a quotation cited by Allen Birchler in "The Anti-Evolutionary Beliefs of William Jennings Bryan," *Nebraska History* 54 (1973), Bryan himself acknowledged the existence of fossils but gave them this interpretation:

> The fossils of extinct animals found in old rock, together with the absence of existing types in those rocks, [make] it reasonable to hold that creation has been a continuous process, new types being created from time to time without any relation to preexisting ones. But there [is] not a bit of evidence that species developed from species. (551)

For Bryan and other creationists, fossils prove periodic creation rather than continuous evolution.

15. William Jennings Bryan, "Man," in *Speeches* 2:298.

16. Ibid., 298.

17. Bryan, *The Bible and Its Enemies*, 31.

18. *The World's Most Famous Court Trial* (Cincinnati: National Book Company, 1925), 323. Further page references to this unedited transcript of the Scopes Trial appear in parentheses in the text.

19. Bryan, "The Menace of Darwinism," 93.

20. Levine summarizes several books that Bryan read linking evolution to Nietzsche, the German military buildup prior to World War I, and the decrease of faith in the Bible among college students (*Defender of the Faith*, 261–265). Bryan also talked with, and received letters from, many college students and their parents about the impact of evolution on their faith. See, for example, Mary Baird Bryan, *The Memoirs of William Jennings Bryan* (Philadelphia: John C. Winston, 1925), 479.

21. George M. Marsden, *Fundamentalism and American Culture: The Shaping of Twentieth-Century Evangelicalism, 1870–1925* (New York: Oxford University Press, 1980), 44.

22. Larson, *Trial and Error*, 49–53.

23. Kenneth K. Bailey, "The Enactment of Tennessee's Anti-Evolution Law," *Journal of Southern History* 16 (1950): 475. This article gives many other interesting details about the origin and passage of the act, including some humorous attacks made on it by college students at Vanderbilt University and the University of Tennessee.

24. Interestingly enough, in the same legislative session a law was passed requiring that public schools not discriminate in their hiring practices against teachers who were atheists or evolutionists (Larson, *Trial and Error*, 58).

25. Quoted in Willard H. Smith, "William Jennings Bryan at Dayton: A View Fifty Years Later," *Proceedings of the Indiana Academy of the Social Sciences*, 3rd ser. 10 (1975): 81. Many other attorneys were involved on both sides of the trial besides these two.

26. Many sources describing the trial are listed in Ferenc Morton Szasz, "The Scopes Trial in Perspective," *Tennessee Historical Quarterly* 30 (1971): 289. The

following treatment of the trial relies primarily on Larson, *Trial and Error*, and on *The World's Most Famous Court Trial*.

27. The issue of the separation of church and state will not be discussed here because it played no major role in this case. Indeed, the act itself un-self-consciously included the word *Bible*. According to Edward Larson (*Trial and Error*), an argument against the act based on the Establishment Clause of the First Amendment could not have been made because such an argument was unavailable within the American legal system at the time (93). This argument became available in U.S. courts only after a 1940 Supreme Court decision affirmed for the first time that a "wall of separation" existed between church and state. This interesting issue is treated at length in Chapter 4.

28. The text for this speech is taken from *The World's Most Famous Court Trial*, an unedited transcript of the Scopes Trial, with page numbers indicated in parentheses. Quotations from this book have been slightly regularized to correct for obvious misspellings and other mechanical and grammatical errors.

29. Histories of the rise of fundamentalism and the role of inerrancy include James Barr, *Fundamentalism* (Philadelphia: Westminster Press, 1978); Ernest Sandeen, *The Roots of Fundamentalism: British and American Millenarianism, 1800–1930* (Grand Rapids, Mich.: Baker Book House, 1970); Marsden, *Fundamentalism and American Culture*; and Szasz, *The Divided Mind of Protestant America*. The theological doctrine was developed in the late nineteenth century by Princeton theologians such as Benjamin B. Warfield.

30. Barr, *Fundamentalism*, 53.

31. Marsden, *Fundamentalism and American Culture*, 122, 61.

32. This speech also appears in *The World's Most Famous Court Trial*. Page references are given in parentheses in the text. Italics are added.

33. He does not necessarily approve; he adds, "If we depended on the agreement of theologians, we would all be infidels" (33).

34. Throughout the cross-examination, Bryan made many other statements backing away from a position of strict literalism, which is often unhelpfully conflated with inerrancy. The difference between these positions is discussed in Barr, *Fundamentalism*, 40–55. For example, Bryan said that God may have used language that could be understood at the time of the event, and that he did not think a day meant a literal twenty-four-hour period, but an indeterminate length of time.

35. The questions and answers appear in Leslie H. Allen, *Bryan and Darrow at Dayton* (New York: Arthur Lee, 1925), 168–169.

36. Some of these other dichotomies and terminological strategies have been discussed by theologians. A brief account of the Bible as a set of representative anecdotes is William C. Placher, *Unapologetic Theology* (Louisville: Westminster/John Knox, 1989), 131. A more detailed treatment of the Bible's function as a narrative appears as Chapter 1 in Stanley Hauerwas's book, *Christian Existence To-*

day: Essays on Church, World, and Living in Between (Durham, N.C.: Labyrinth Press, 1988).

37. One good essay on belief is William James, "The Will to Believe," *Essays in Pragmatism* (New York: Hafner, 1948), 88–109. *Belief* is also an important term in the position known as neopragmatism. See *Against Theory: Literary Studies and the New Pragmatism*, ed. W.J.T. Mitchell (Chicago: University of Chicago Press, 1985).

38. Barr, *Fundamentalism*, 40.

39. See Ronald L. Numbers, "The Creationists," in *God and Nature: Historical Essays on the Encounter between Christianity and Science*, ed. David C. Lindberg and Ronald L. Numbers (Berkeley: University of California Press, 1986), 402–403; and George M. Marsden, "A Case of the Excluded Middle: Creation versus Evolution in America," in *Uncivil Religion: Interreligious Hostility in America*, ed. Robert N. Bellah and Frederick E. Greenspan (New York: Crossroads Press, 1987), 145–146. Both of these long articles summarizing Bryan's role in the movement have also been reprinted in shorter form, "Creatism in 20th-Century America," in *Science* 218 [1982]: 538–544, and "Creation versus Evolution: No Middle Way," *Nature* 305 (1983): 571–574. Both authors attack Bryan's motives more than they make sense of the limits of his literalism.

40. Daniel J. Boorstin, *The Image: A Guide to Pseudo-Events in America* (New York: Atheneum, 1961, 1985).

41. Ashby provides many interesting details about Bryan's own studies of rhetoric, including his extended education in Latin (three years) and Greek (four years) (10), his practice at speaking with pebbles in his mouth (35), his first experience with his own oratorical power over "a noisy small-town crowd" (he told his wife afterwards "I could move them as I chose" and prayed that he would never misuse his talent) (26), and his characteristic stage fright before speaking (50). Bryan also acted as editor-in-chief for a ten-volume series entitled *The World's Most Famous Orations* (New York: Funk and Wagnalls, 1906), for which he wrote a preface in which he explains his view of rhetoric and comments on his favorite speeches and strategies.

42. Bryan's post-trial speech uses the same arguments discussed in relation to his previous antievolution speeches. It has been collected as an appendix to the biography by his wife (Bryan, *Memoirs*) and is also reproduced in many books about the Scopes Trial.

43. Smith, "Bryan at Dayton," 81.

44. Clarence Darrow, *The Story of My Life* (New York: Charles Scribner's Sons, 1932). Subsequent page references appear in parentheses in the text.

45. Quoted in Levine, *Defender of the Faith*, 40.

46. Ibid., 166.

47. Ashby, *William Jennings Bryan*, 51.

48. Ibid., 144–145.

49. Ibid., 64–66.

50. Quoted in Levine, *Defender of the Faith*, 357.

51. Mary Baird Bryan, *Memoirs*, 487.

52. Quoted in Levine, *Defender of the Faith*, 357.

53. H. L. Mencken, *The Impossible H. L. Mencken: A Selection of His Best Newspaper Stories*, ed. Marion Elizabeth Rodgers (New York: Doubleday/Anchor Books, 1991), 592. Further page references appear in parentheses in the text.

54. H. L. Mencken, "In Memoriam: W.J.B.," in *Prejudices: Fifth Series* (New York: Alfred A. Knopf, 1926), 64. Further page references appear in parentheses in the text.

55. The nickname the Scopes "Monkey" Trial was first coined by Mencken himself. The metaphor of the human as a monkey is not only an important trope in the creation/evolution controversy (as was discussed in the previous chapter in relation to the Huxley/Wilberforce debate); it is also a crucial figure of speech in many other public discussions both before and after Darwin, including most discussions of race in the United States. For a study of the rhetoric of race in America, see Henry Louis Gates, Jr., *The Signifying Monkey: A Theory of Afro-American Literary Criticism* (New York: Oxford University Press, 1988). The powerful metaphor of human as monkey deserves further study.

56. Arthur Garfield Hayes, "The Scopes Trial," in *Evolution and Religion: The Conflict between Science and Theology in Modern America*, ed. Gail Kennedy (Boston: D. C. Heath, 1957), 35–52; and Kirtley F. Mather, "The Scopes Trial and Its Aftermath," *Journal of the Tennessee Academy of Science* 52 (1982): 2–9.

57. Two other examples deserve brief mention, the first because it links Bryan to rhetoric, and the second because it reveals the scientists' own religious vocabulary. In an essay entitled "The Christian Statesman" in *American Mercury* 3 (1924): 385–398, Edgar Lee Masters (author of the *Spoon River Anthology*) vilifies Bryan and explains his problem in these terms: "A poet tries to tell the truth; an orator tries to persuade. Bryan was always bewitched with the art of persuasion" (393). Masters shows that in the early twentieth century, not only American scientists but also American poets tried to define themselves in opposition to rhetoric so as to avoid contamination by human perspectives and beliefs.

The other example is Maynard Shipley's 400-page book, *The War on Modern Science: A Short History of Fundamentalist Attacks on Evolution and Modernism* (New York: Alfred A. Knopf, 1927). The final chapter of this massive work, entitled "What Is to Be Done?" ends with the claim that unless Americans work to stop the creationists, "we, the people, [will be] preparing for the crucifixion of science, the savior of the world" (384). This book, by an evolutionist and the president of the Science League of America, does exactly what the creationists fear: it posits a new savior in science itself.

58. Walter Lippmann, *American Inquisitors: A Commentary on Dayton and Chicago* (New York: Macmillan, 1928), 6. Further page references appear in the text.

59. I take the terminology battle as more basic than Lippmann (ibid.) suggests; there is, finally, no stable relationship between *reason* and *freedom* except as defined by each side.

60. Various accounts of the trial and treatments of Bryan are summarized in Willard Smith, "Bryan at Dayton"; Larson, *Trial and Error* (72–75); and Szasz, "The Scopes Trial in Perspective." A full bibliographic essay discussing many sources is William E. Ellis, "Evolutionism, Fundamentalism, and the Historians: A Historiographical Review," *Historian* 44 (1981): 15–35.

61. Szasz, *Divided Mind of Protestant America*, 123.

62. Judith V. Grabiner and Peter D. Miller, "Effects of the Scopes Trial," *Science* 85 (1974): 832–837; and Robert M. May, "Creation, Evolution, and High School Texts," in *Science and Creationism*, ed. Ashley Montagu (New York: Oxford University Press, 1984), 306–310. There is no shortage of opinions about what this particular fact portends for education and democracy.

63. Larson, *Trial and Error*, 75.

64. Murray Shumach, " 'Monkey Trial' Staged," *New York Times*, 21 April 1955, II:3.

65. Ibid.

66. Jerome Lawrence and Robert E. Lee, *Inherit the Wind* (New York: Bantam, 1955), ix. In the analysis that follows, page references in parentheses specify quotations from the play. Where no page reference occurs, the quotation comes from the movie version and was either added to, or revised from, the play.

67. Aside from numerous newspaper reviews, only three brief literary critiques of the movie and the play have been published: a source study, a character analysis, and an analysis of the genre of the work. These studies are, respectively, Michael P. Dean, "Language and Character Formation in *Inherit the Wind*," *Publications of the Arkansas Philological Association* 8 (1982): 16–23; Susan Duffy, "The Origin of Speeches: *Inherit the Wind*, Irving Stone, and the Scopes Trial," *American Notes and Queries* 22 (1983): 14–17; and Allen E. Hye, "A Tennessee Morality Play: Notes on *Inherit the Wind*," *Markham Review* 9 (1979): 17–20. None of these studies attempts a rhetorical analysis of the type developed here.

68. This subtle reference to the Ku Klux Klan suggests an important link in this work between creationism and southern racism. Apparently with his suggestions about monkeys, Darwin threatened not only the difference between humans and animals, but the much more tenuous difference between blacks and whites. Throughout American history up to the time of the civil rights movement, the metaphor of the black person as a monkey was crucial (as was suggested in Note 55).

The opposition between the North and the South implied by this allusion is another crucial dichotomy for the creation/evolution controversy as a whole. I have treated this dichotomy in detail in an essay revised from this chapter: "Attacking (Southern) Creationists," in *Religion in the Contemporary South: Diversity, Community, and Identity*, ed. O. Kendall White, Jr., and Daryl White (Athens: University of Georgia Press, 1995), 59–66. In brief, I argue that the North/South dichotomy is regularly linked to other dichotomies such as city/country, educated/uneducated, and progressive/backward so as to privilege evolutionism as a truth that is widely accepted in the urban North and to devalue creationism as a belief held and acted on only in the rural South. In reality, although the 1920s creationist laws were passed mostly in the South, antievolution bills were also introduced in several northern and western states, and George Marsden points out that fundamentalism as a whole has largely been an urban and northern phenomenon (*Fundamentalism and American Culture*, 188). Such facts suggest that the North/South opposition has been used by many people to assign reason and belief to different geographical areas.

69. This implication is supported in the play (and even more in the movie) by abundant references to Socrates, Copernicus, Galileo, hemlock, and lynching. This work regularly represents the Scopes Trial as a repetition of these earlier trials, which are similarly narrated in our culture as contests between scientific rationality and religious belief.

In contrast to this representation of the townspeople in the work, both Mencken and Darrow reported that the people of Dayton were very warm and friendly to them. Darrow wrote that, despite a severe shortage of dairy goods and ice due to the size of the crowds, some anonymous residents left for him, in the icebox of his borrowed house, "a slab of ice, [some] milk and cream and butter, and even a choice cantaloupe for Monday breakfast." Darrow felt that in Dayton he had encountered "true Southern hospitality" (*Story of My Life*, 252). Mencken wrote in an early news report: "I expected to find a squalid Southern town. . . . What I found was a country town full of charm and even beauty" in which "the Evolutionists and the Anti-Evolutionists seem to be on the best of terms and it is hard in a group to distinguish one from the other." In regard to the treatment of Scopes, Mencken writes that the Tennessee officials guaranteed "no one [would] be permitted to pull his nose, pray publicly for his condemnation, or even to make a face at him" (*The Impossible H. L. Mencken*, 569). These details, which were chosen for different rhetorical purposes, evoke a picture of kindly Southerners, which contrasts significantly with the picture of a lynch mob evoked by the movie.

70. This distinction between the language of law and the language of politics and religion is very curious given that many of the ancient rhetorics were handbooks for lawyers and that lawyers have usually been thought of in Western culture as using many language tricks. *Inherit the Wind* contains several playful attacks on lawyers and journalists as well as on creationists, but it finally portrays the lawyer Drummond as the most powerful advocate for truth and the only person who can silence Brady's speaking by identifying it as rhetoric.

71. Later in the movie, Hornbeck gives an alternative explanation of heaven as a commodity. He says that when people saw the stars, they "decided they were groceries belonging to a bigger creature. That's how Jehovah was born."

72. As Rachel's simile suggests, the movie extensively correlates the difference between childhood and adulthood with the difference between belief and thought. Rachel's own rejection of creationism is depicted as a growth from child-

hood to adulthood, in which she gradually leaves behind her natural father and her foster father, Brady. In contrast, Brady's refusal to give up his belief in creationism is portrayed as a return to the state of infancy. In many other attacks on organized religion besides those in this movie, the metaphor of human growth is also used for rhetorical effect. In response to Rachel's comparison of a thought to a child, a creationist might ask what a belief is like and what happens to the believer when it dies. A common answer is that the believer knows the belief was not true, and so ought to feel relieved. Such a conception reveals again the effects of the truth/belief dichotomy in Western culture.

73. This song is usually sung by a group of intimidating women, who may represent a cultural fear of women attaining new political power (and the right to vote) during the 1920s through the efforts of men like Bryan. Women are similarly depicted as threats in many western movies.

74. This latter song suggests again the North/South conflict implicit in this controversy. It was written by an abolitionist (Julia Ward Howe) to suggest that God would take the side of the North against the South during the Civil War. Many Southerners still hate it.

75. Stephen Jay Gould, "A Visit to Dayton," in *Hen's Teeth and Horse's Toes* (New York: Norton, 1983), 273. Subsequent page references appear in parentheses in the text.

76. Later in the essay Gould asks a rhetorical question: "Can anyone take seriously a link between Darwinism and the four evil p's?" (ibid., 277). Creationism itself suggests that the answer is yes.

77. The role of these faulty definitions and nonsense words in the 1980s controversy will be examined in Chapter 4.

4

The Arkansas
Creation-Science Trial

The most famous contemporary episode of the creation/evolution controversy began in early 1981 when the state of Arkansas considered and enacted Act 590, the Balanced Treatment for Creation-Science and
Evolution-Science Act. This law required that creationism be taught in Arkansas public schools whenever evolution is taught. In May 1981 the
American Civil Liberties Union (ACLU) filed a civil suit against the law
on behalf of the Reverend Bill McLean, an Arkansas Presbyterian leader,
and twenty-two other plaintiffs, including parents, teachers, and several religious, scientific, and educational organizations.[1] The plaintiffs requested
that the law be overturned because it violated the separation of church and
state that is mandated by the First Amendment to the U.S. Constitution.

The trial, *McLean versus Arkansas Board of Education* (*McLean v. Arkansas* for short), was held December 7–17, 1981, in Little Rock before Judge
William R. Overton of the U.S. District Court for Eastern Arkansas. Participants included the New York law firm of Skadden, Arps, Slate, Maegher,
and Flom, plus two local law firms, representing the plaintiffs; the Arkansas
Attorney General's Office, representing the defendant; sixteen witnesses
for the plaintiffs; and eleven witnesses for the defense. Although the trial
was immediately dubbed "Scopes II" by the media, in contrast to the Scopes
Trial where expert witnesses were excluded from the case, witnesses in the
Arkansas Trial were mostly professors from American universities who were
called as experts on their respective disciplines. Taking notes on the trial
and drawing sketches were reporters representing seventy-one news organizations from around the world, who sat in the unused jury box (a location

rich in symbolic implications). On January 5, 1982, Judge Overton delivered a thirty-eight-page decision finding for the plaintiffs and overturning the law on grounds that it violated the U.S. Constitution. Thus ended the best-known recent creation/evolution episode, in what appeared to be a decisive defeat for the creationists.

However, in the months after the decision, at least six books appeared by scientists attacking creationism.[2] The authors explained in their prefaces that these books were needed because the creationists refused to give up their fight even though they had just lost a major court case. In fact, the Mississippi legislature passed a bill requiring balanced treatment for creationism the day after Judge Overton's ruling, and similar bills remained before the legislatures of at least seventeen other states. These bills convinced scientists and science teachers around the nation that creationism continued to pose a serious threat to evolution. Consequently, they decided to find other ways besides the trial to persuade the nation that it must repudiate creationism and keep it out of the public schools.

These scientists were joined by other academics, especially the expert witnesses who appeared in the trial, who were also persuaded that they must join the fight against creationism and subsequently published at least a hundred articles in academic journals giving their own interpretations of the creationist phenomenon. Special issues were entirely devoted to the creation/evolution controversy.[3] A new journal, *Creation/Evolution*, was founded to give evolutionists a chance to "work together and pool their efforts just as the creationists do."[4] Essays were written and compiled into at least ten collections by anti-creationist scientists.[5] A special course about the controversy was taught at Iowa State University.[6] A collection of condemnatory reviews of creationist books was compiled.[7] Popular magazines and public television stations produced articles and programs dealing with the controversy.[8] Bibliographies were assembled.[9] An anti-creation-science dictionary was printed.[10] Some participants in the trial even wrote personal narratives of their involvement which resemble mystery novels; they tell an exciting story and reveal "whodunit" in a trial that apparently was as exhilarating and dangerous as a murder case.[11] Why was a 1981 civil case in Arkansas able to become such a rich intellectual issue, to pose such a threat, and to arouse such a degree of anxiety and interest? Why did journalists, magazine editors, and television commentators focus so much attention on this trial? Why did it overshadow a much more important creationism case: the 1988 decision by the U.S. Supreme Court, the highest judicial body, to overturn the Louisiana Balanced-Treatment Act which was similar to Arkansas Act 590?[12] Why did academics in fields as diverse as biology, education, history, law, philosophy, sociology, and theology make such significant efforts to combat creationism and defend evolution?

This chapter attempts to answer these questions by presenting a close reading of the trial and many of the academic texts it spawned. It begins by picking up the rhetoric/philosophy thread that is the focus of this volume. After tracing the key terms of the controversy from 1925 to 1981, it reenacts the trial itself by analyzing the testimonies given and works about creationism subsequently written by trial participants from different disciplines. It concludes by looking at the text of Judge Overton's decision to see why he was persuaded that creation-science was unconstitutional. It argues that this episode of the creation/evolution controversy was another landmark terminology battle, just like the debate about Darwin's *Origin of Species*, the Huxley/Wilberforce debate, and the Scopes "Monkey" Trial. On one side were the creationists, committed to a literal reading of the biblical account of creation and opposed to a public school system that they think teaches children to deny their faith. On the other side were academics from many disciplines, each with their own disciplinary commitments to particular definitions of terms but sharing an overall commitment to human reasoning based on verifiable evidence and skeptical doubt and to a political philosophy that tolerates private religious beliefs by distinguishing them from public knowledge. I see in this episode of the controversy a compelling example of the great contemporary battle raging throughout much of the world between humanists and fundamentalists. In telling their own stories of the trial, both sides attempted to persuade the culture as a whole to accept their terms for the controversy and to enact their position as public policy.

RHETORIC VERSUS PHILOSOPHY

Of the six anticreationist books published immediately after the Arkansas Trial, the most highly acclaimed was *Abusing Science: The Case against Creationism*, by philosopher of science Philip Kitcher. Widely reviewed by other scientists and philosophers, this book was called "an impressive and absorbing counterattack" on creationism, and "an ideal reading for a course in the philosophy of science" that deserves "frequent re-reading" and strikes an excellent balance between the need to "address the specific issues that creationists raise" and "emphasize their problems of methodology, logic, and [proof]."[13] This generous praise suggests that the book exemplifies the evolutionists' reactions to creationism.

As an exemplary reaction the book gives two central clues to the controversy as a whole. First, its primary goal is to prove that creation-science is not science but "an abuse of science"; it assumes that one can resolve the terminology battle by determining the proper definition of *science*, and measuring the opposing positions against this definition. Second, it implies that creationists reject the correct definition of *science*, not because they disagree with it, but because they do not understand it or refuse to accept it; they deny the truth of evolution because they are committed to a dangerous political agenda and to untenable religious beliefs. By posing the conflict as a struggle between scientific knowledge, on the one hand, and politics and religion, on the other, the book reenacts the recent conflict that is traced throughout this volume and involves these versions of rhetoric and positivist philosophy.

In his introduction Kitcher begins by launching a direct attack on rhetoric. Termed "The Creationist Crusade" in an allusion to the Crusades of the Middle Ages, the introduction begins by narrating in a paragraph the story of the Huxley/Wilberforce debate. In summarizing this mythic encounter, Kitcher argues that Wilberforce, "a skilled debater whose slippery performances had earned him the nickname of 'Soapy Sam' . . . thought he saw a way to achieve rhetorical effect" by asking Huxley which of his grandparents was an ape, but that Huxley "went on to deliver a scathing response, openly admitting that he would prefer an ape for a grandparent to a man, blessed with intellect and education, who used rhetorical tricks to confuse an important scientific issue."[14] Kitcher links the term *rhetoric* to Wilberforce twice in this paragraph, implicitly contrasting Wilberforce's rhetoric to the simple truth delivered by Huxley. In his second paragraph, he connects this 1860 story of the deceptions of rhetoric to the present day: "Over 120 years later, the conclusions and debating methods of Soapy Sam are alive and well and playing in Peoria" (1). He asserts that the 1980s creationists were using similar language tricks to defend creationism and attack evolution.[15]

Later in the introduction, Kitcher applies his critique of rhetoric to the Arkansas trial, which had already been decided before his book went to press. He explains why the book is still needed:

When the Arkansas Creationist law was challenged, a team of scientific, philosophical, and theological *luminaries* assembled in Little Rock. Judge Overton drew on expert testimony. Teachers, administrators, and local school-board members [in other towns] are likely to be less *lucky*. The advocates of Creationism may be *winsome, tactful, and sweetly reasonable*, but they are bent on having their way and they are equipped with the ideas and arguments of the Institute for Creation Research. Ready with the *rhetorical tricks* of Creationist Literature, a carefully primed student or a concerned parent can easily embarrass a teacher or a PTA [Parent-Teacher Association] member. Not every school district has its Huxley, prepared to respond to the *clever questions* of the local Wilberforce. (3–4)

This quotation indicates that although the Arkansas judge was able to defeat creationism by relying on expert testimony from academic "luminaries," people in local schools would not be so "lucky." Instead, they would have to face the "winsome, tactful, and sweetly reasonable" creationists alone. In order to help the local teachers and PTA members avoid "embarrassment" from the creationists' "rhetorical tricks" and "clever questions," Philip Kitcher, an expert philosopher of science, wrote this book to provide the knowledge needed to counter their rhetoric.

Kitcher concludes his introduction by overtly linking rhetoric to politics and religion. He argues that his book is not "an attempt to debunk religion," nor even to attack "a literal reading of the Genesis account of Creation simply as an article of religious faith."[16] Rather, he indicates, "my business is strictly with a political movement," creationism, which deliberately attempts to discount the "scientific evidence" that "tells decisively against the literal truth of Genesis." He concludes, "I quarrel only with those who *pretend* that there is scientific evidence to favor the Genesis story understood literally, who *masquerade* religious doctrines as scientific explanations, and who try to *persuade* their fellow citizens to make religious teaching a part of education in science" (6). The key verbs of this passage are *pretend, masquerade, persuade;* they are linked to the devalued member of the key opposition of *religion* versus *science.* The passage implies that science delivers sincerity rather than pretenses, the naked idea rather than the costumed sophism, a demonstration of the truth rather than persuasion to a belief or opinion. In this introduction with its three attacks on rhetoric, Kitcher argues that the creationists could only defend their position by lying. He is willing to let them retain their literalist beliefs in the Genesis story but does not want them to mistake these beliefs for scientific truths like evolution.

Kitcher is not the only evolutionist to see in this episode of the creation/evolution controversy a contemporary battle between rhetoric and philosophy. In his powerful, well-argued book *Science on Trial,* evolutionary biologist Douglas Futuyma attacks rhetoric more explicitly.[17] Comparing creationism to a fortress (and thus continuing the image of creationism as a last holdout from the Middle Ages), Futuyma indicates what holds this fortress together: "The cement is, for the most part, rhetoric, the tool of the Sophists, who taught their pupils how to win arguments, rather than how to seek for truth" (178). His book-length attack on rhetoric concludes:

But what reason has achieved, it has achieved in the face of the opposition of authority and tradition. If authority had its way, we still would not know what Galileo saw through his telescope. It is inherent in rationalism, and in the apotheosis of rationalism that is science, that tradition and authority must give way to new worlds of thought. The voice of authority must be forever threatened by the rational mind that dares to doubt, and so rationalism is our best defense against political tyranny. Authority draws its strength from its reliance on law and force; rationalism from its faith in the human mind. (221)

Futuyma sees in this controversy a battle pitting reason against authority and tradition, the modern world against the Middle Ages, Galileo against the church, law and force against the human mind. He reveals his own key values and commitments by defending rationalism and "the apotheosis of rationalism that is science," positions that put their "faith in the human mind" rather than in tradition and authority. Ironically, the book itself is a rhetorical effort inasmuch as it is a defense of Futuyma's faith; in it he uses his own authority within the tradition of evolutionary biology to argue for evolution against creationism, for truth and reason against rhetoric and belief. If the truth of evolution were obvious and unmistakable to everyone, he would not have needed to write this book.

Not only the evolutionists, but the creationists as well, see this controversy as a battle between rhetoric and philosophy. In his preface to a book about the Arkansas trial, Norman Geisler, a theologian and defense witness, touches on the same central theme as Kitcher. Geisler argues that he must counter "the slanted and distorted reports [of the trial] in the secular press" by relating "the truth of what happened" in a book that "is almost entirely documentary."[18] He wants to replace the rhetoric of the news media with a true and direct account. His attitude toward rhetoric is echoed by Duane Gish, one of the leaders of the contemporary creationist movement, who writes in a foreword to Geisler's book that it is a "refreshing contrast to the usual (not always) distorted accounts which appeared in the mass media and a relief from the sophistry that appeared in so many scientific journals."[19] By using the word *sophistry*, the most common synonym for *rhetoric* in the Western tradition, Gish suggests that the truth of the Arkansas Trial must be told to counter the lies spread by the evolutionists. Both sides of this controversy argue that the truth of philosophy must overcome the deceptions of rhetoric, and both agree that one ought to use straightforward language to tell the plain facts about humanity's origin.

The problem is that both sides see different plain facts about the question of our origin. By claiming to tell the naked truth revealed in simple language rather than lies dressed in the costume of rhetoric, both sides focus on the most important recurring dichotomy in each episode of the creation/evolution controversy: truth/error. The episodes analyzed in this study cumulatively suggest that the word *truth* is perhaps the most important rhetorical tool in a post-Enlightenment culture; it is the word that best persuades others to accept one's own values and beliefs. When someone asks, "Why should I believe you?" the strongest possible answer in such a culture is, "Because I am telling you the truth." People who use this phrase are not lying; they are reporting what seems obvious and unmistakable to them. My key argument is that this obviousness and unmistakability is not a function of some match between word and thing, assertion and reality. Rather, it is a function of a particular worldview developed within a particular community of people who use words as imperfect tools to work together on the basis of shared perceptions and commitments. What does one do when members of other communities fail to share these perceptions and commitments? In our culture, one accuses them of lying. After all, the assumption goes, one's key words must have obvious meanings, which one's opponents simply refuse to concede.

As with the debate about Darwin's *Origin of Species*, the Huxley/Wilberforce debate, and the Scopes "Monkey" Trial, the Arkansas trial did not focus directly on the meanings of the words *truth* and *error* but rather on the parallel dichotomy science/religion, which has become increasingly important in this debate since the publication of Darwin's *Origin*. In 1947 the U.S. Supreme Court made a decision that shifted these terms in a significant way and effectively banished the term *religion* from this public discussion. This decision interfered with the dialogue between creationists and evolutionists even as their disagreements intensified. Instead of seeing different human communities in disagreement on basic values and beliefs, as encoded in complex linguistic signs, both sides continued to see a battle between rational people on their own side who reveal obvious truths and irrational people on the other side who propagate lies and use rhetoric to conceal them.

KEY TERMS OF THE CONTROVERSY
FROM 1925 TO 1981

Although the Scopes Trial was subsequently mythologized into a resounding defeat for the creationists, it was not taken this way by most contemporary observers. In fact, in the years immediately following the Scopes decision, several other states passed antievolution laws, including—significantly—Arkansas, which passed such a law in 1928. Unconvinced of the truth of evolution, creationists continued to work to outlaw it in many other states until about 1930, when they mysteriously quit. Some historians have suggested that the Depression diverted their attention to more urgent matters. Others have suggested that high school biology textbooks stopped disseminating evolution, and thus that the reason to enact such laws disappeared. A few researchers have claimed that the fundamentalists gradually diminished in power as an ever-larger percentage of the population came to accept the authority of science itself. No matter what the explanation (each of which could be analyzed in terms of its own political and religious commitments), it is clear that many schools did not systematically begin to teach evolution until another dramatic event took place.

The triggering event was the launching of *Sputnik* I by the Soviet Union in 1957. When this first manned space satellite returned safely to earth, many American scientists and politicians decided that the United States must catch up to the Soviet Union or else face the possibility of annihilation. Among other measures, these leaders convinced the U.S. National Science Foundation to fund a new project for $7 million named the Biological Sciences Curriculum Study (BSCS); its mission was to help American students catch up with the Russians in the biological sciences. The BSCS soon decided that American students could catch up only if high school science textbooks taught evolution as the central focus of biology.[20] At the National Science Foundation Summer Institute in 1959, a Darwin centennial was held to commemorate the publication of the *Origin of Species*; the keynote address was entitled, "One Hundred Years without Darwin Are Enough."[21] Taking advantage of this climate of opinion and the new and mutually beneficial collaboration between the federal government and professional biologists, BSCS professors produced three new biology texts. These texts were eagerly accepted by three large publishing houses, who trusted in the power of federal support to assure vigorous sales.[22]

As a result of BSCS influence, many science classes throughout the country began to teach evolution again. Ultimately, nearly half of American high schools adopted the BSCS texts (along with supporting filmstrips, periodicals, and other materials). This movement affected the rest of biology texts as well; the other publishers had to work to update their books in order to compete in a free market with the government-sponsored editions.[23] Before long, many students were learning about the theory of evolution, which some parents took as a denial of the truth of Genesis and the existence of God. As a result, a number of creationists began antievolution initiatives, especially in Texas and California.

Perhaps the most significant event in the recent development of American creationism was the publication in 1961 of *The Genesis Flood*, a 500-page book written by a professor of the Old Testament, John C. Whitcomb, and a professor of civil engineering, Henry M. Morris.[24] In this book Whitcomb and Morris argued that the scientific evidence cited by evolutionists could be interpreted more cogently as proof of the details described in Genesis. The authors admitted outright that they had first believed literally in the Genesis account and then had begun to assemble evidence from science to support it; their first phrase asserts that they had labored "[i]n harmony with our conviction that the Bible is the infallible word of God."[25] The book thus did not claim to create a new science that coincidentally matched their religious beliefs, but it did found a new creationist approach: the assembling of scientific evidence for creation and against evolution. In 1974 this evidence was summarized and the movement was renamed by a book entitled *Scientific Creationism* and written by the same Henry M. Morris.[26] Within the next few years, contemporary American creationism was renamed again—*creation-science*—so as to emphasize its claim to scientific status; this is the term that appeared in the 1981 Arkansas act that was overturned by Judge Overton. Creation-science, the 1970s and 1980s version of creationism was, finally, itself a reaction to politics in the form of a U.S. government mandate that widely propagated evolution as a partial cure for our national inferiority to the Russians. Given this context, *Inherit the Wind* could be read even more closely than in Chapter 3, as a Cold War story about some irrational beliefs that threatened U.S. national security.

In the years after the Scopes Trial, there were changes in other key terms for this controversy besides *creationism*. The key term *religion* also shifted in significant ways, especially because of a reinterpretation of the First Amendment as applied to public schools. In his essay, "Religion in the Schools: A Historical and Legal Perspective," political scientist Donald E. Boles traces the history of public education in the United States from the time of the Puritans to the Arkansas trial.[27] In brief, Boles argues that religious sects actually founded public education in America, primarily as a way to teach students to read the Bible and to become good citizens. However, as the early Protestant majority lessened in the face of late-nineteenth-century immigration, the newly arrived Catholics, Jews, and other religious minorities began to object to the sectarian impulses in the schools, and as a result, the courts, in a series of church-state cases, gradually reduced the influence of religion. The earliest of these was *Reynolds v. United States* (1880), a case in which the Supreme Court ruled that Mormon polygamists could believe in polygamy but could not practice it. Such cases have continued until the present day. Throughout this century of rulings, the Court has made an effort to be faithful to both prongs of the First Amendment, which reads: "Congress shall pass no law respecting an establishment of religion or prohibiting the free exercise thereof."

In trying to determine the meaning of these two clauses, as well as to deal with the notoriously ambiguous word, *or*, that links them, judges have predictably followed the time-honored tradition of going back to the intent of the framers, which, according to Boles, clearly revealed "a deep suspicion of organized religion, particularly about the effects it might have on education." In support of his reading, he paraphrases "Jefferson's leading biographers" that Jefferson "believed the First Amendment would create a high and impregnable wall of separation between church and state."[28] However, according to Edward Larson, the Supreme Court did not apply this "wall of separation" metaphor to the First Amendment until 1947, when Justice Hugo Black used this phrase for the first time in American legal history—in the second draft of his decision for the case *Everson v. Board of Education* (330 U.S. 1) (which held that parents can be reimbursed from public funds for the costs of transporting their children to parochial schools). Larson argues that this metaphor actually countered the standard nineteenth-century reading of the First Amendment (which he illustrates with an 1833 quote from Justice Joseph Story) which asserted that the amendment only prevented the federal government from sponsoring a particular church.[29] In fact, Larson reports that many states had their own state-sponsored churches during the first fifty years after the American Revolution.[30] In Larson's view, the framers did not create a strict separation of church and state; their successors did so, long after the framers had failed (as everyone inevitably must) to make their original intent unmistakable.

Another insightful reading of the First Amendment has been suggested by John F. Wilson, a professor of religion at Princeton University. In a lecture delivered at Duke University in 1989, Wilson explained what he had found in his study of documents relating to the First Amendment from the Constitutional Convention.[31] In investigating what kinds of arguments had surrounded the drafting and ratification of this amendment, he discovered to his astonishment that there were none. What the amendment meant was apparently clear and agreeable enough to all the delegates that they simply rubber-stamped it. As a result of further research, Wilson concluded that the framers probably saw the freedom of religion as the freedom to choose between churches, not the freedom to ignore churches altogether; at that time virtually all the delegates belonged to one church or another and did not want the federal government telling them which one to join or to support with their tax dollars, a practice common in many of the states at the time of the Revolution. Wilson's discovery can be profitably compared to Michael Buckley's thesis in At the Origins of Modern Atheism; apparently, the framers did not think of atheism as a serious possibility in the same way that twentieth-century people do, and thus it did not occur to them that the First Amendment might be used to devalue religion as a whole in favor of nonreligion. Consequently, Wilson concludes that it is pointless to ask what the framers originally intended on an issue that they could not have possibly considered.

Still attempting to determine what the term *religion* meant to the framers, Wilson then decided to trace the wall-of-separation metaphor to its original source. Expecting to find it in the Constitutional Convention itself, he was amazed to discover that it first appeared in print fifteen years after the First Amendment had been ratified, in a private letter that Jefferson had written. This metaphor was never a part of the official discussions in any state during or after the ratification process (Jefferson himself was in France at the time); nor did it seem to reflect Jefferson's intent at the time of the convention. It was apparently discovered after Jefferson's death by his biographers, who used it because it captured their own imaginations and provided a vivid image with which to argue their own interpretations of Jefferson's attitude toward religion. These sources introduced the metaphor to people like Justice Black. As atheism and non-Judeo-Christian religions became more prevalent in twentieth-century America, the Supreme Court needed some basis on which to determine how to treat these other religions (while wondering whether the word *religion* should even be used to describe them). Consequently, the Court looked for someone sufficiently distinguished and relevant to cite as a precedent. Jefferson was an obvious choice, given his importance in the founding of the republic and the imaginative power of his wall metaphor. This metaphor was much clearer and more compelling than anything Jefferson or the other founders originally wrote. Eventually, this metaphor became the governing motif in Judge Overton's decision; the decision literally ends with an allusion to Robert Frost's poem, "Mending Wall." How interesting is the meandering story of persuasion through history!

When Justice Black cited the wall metaphor in the crucial *Everson* case, he thus introduced as a precedent from Jefferson a different interpretation of the First Amendment than had held sway throughout the previous history of the Supreme Court. This interpretation has been truly seminal throughout subsequent legal history, having served as the basis for a host of church-state decisions made since 1947, when the First Amendment began to be legally troublesome almost for the first time since 1776.[32] One of the most important results of this new interpretation was *Engel v. Vitale* (370 U.S. 421), a landmark 1962 case that banned public school prayers and is generally recognized as the first case to exclude a religious practice from the schools. There have been many more cases since then. A more directly related example is *Epperson v. Arkansas* (393 U.S. 97), a 1968 case that held the 1928 Arkansas law banning evolution unconstitutional and thus necessitated a major shift in creationist strategy. Like John T. Scopes, Susan Epperson was a high school biology teacher. While teaching in the Little Rock public school system, she decided that it was unfair for the state to forbid her to teach evolution. Like Scopes, she became a test case for an antievolution law; but unlike Scopes, she failed to become an American legend, perhaps because she lost Scopes's aura of tragedy by appealing to the U.S. Supreme Court and winning the case! Almost nothing has been written about Epperson, the trial, or its outcome, except for brief references in law books.

This survey of First Amendment legal history suggests that when the creationists began to resist BSCS efforts to introduce evolution into public schools, they entered a public arena in which the term *religion* became ambiguous and hotly contested and also open to automatic attack because it appeared in the First Amendment. Motivated by their religious beliefs to contest this policy of federal support for evolution, the creationists were, paradoxically, unable to use the word *religion* in their explanation unless they wanted to lose the contest immediately. As a result, they were left in a difficult rhetorical situation. They wanted to persuade the nation that their religious freedoms were being violated by the teaching of evolution, but they could not use the word *religion* to describe the violated beliefs. They were fighting against an idea that was supported by the government as an essential ingredient in the scientific education necessary for the national defense. Moreover, they were required by the legal system to defend their position on a rational basis in a society that conceived of *reason* as a public and objective process that differed in important ways from the personal and subjective process known as *belief*. How could they argue that their beliefs were not subjective and personal, but rather, compelling and true?

In answer to this question, creationists came up with a new rhetorical strategy: they created an approach called *creation-science*. This term allowed them to defend their beliefs as scientific and thus to take advantage of the prestige granted to the term, and it simultaneously permitted them to get around the serious difficulties attendant on any argument that used the devalued term *religion*. For the creationists, this terminological strategy was not a deception but rather a necessity if they wanted their case to be heard at all. Thus, they responded to Justice Black and his 1947 wall of separation between church and state by inventing a new rhetorical tool.[33]

AN OUTLINE OF KEY ARGUMENTS IN THE TRIAL

This rhetorical tool and all the corresponding tools of evolutionists became the focus of the Arkansas trial in 1981. Indeed, the trial occurred because of the power of the term *creation-science*. The law on trial, Arkansas Act 590, was passed because creationists had successfully argued that a body of knowledge called *creation-science* had been unfairly excluded from public education by science teachers. The law mandated that this knowledge be given a fair place in the curriculum; thus it was called "the Balanced Treatment for Creation-Science and Evolution-Science Act. Because the court case was brought against the law, the plaintiffs first made their case against creation-science.

The plaintiffs' primary strategy was to define *creation-science* not as science but as religion so that it could be excluded from the public schools under the Establishment Clause. In direct opposition to the Scopes Trial, which excluded expert witnesses, the plaintiffs called three distinct teams of expert witnesses, representing religion, science, and education, to show that all three relevant groups of professionals were united in their opposition to creation-science. Each witness testified that within his or her own discipline, creation-science did not qualify as science but rather was considered a religion. The plaintiffs believed that if they could uphold these definitions of terms, they would win the case.

After the plaintiffs had concluded, the Arkansas Attorney General's Office proceeded with its defense. The strategy was to prove that creation-science was indeed science rather than religion and that its inclusion in the curriculum was mandated by American political ideals rather than prohibited by them. Their witnesses argued that both evolution and creation are equally scientific, that creationism better describes certain scientific phenomena than evolution, that professional biologists have excluded creation-science for unscientific reasons, and that therefore, creation-science ought to be included in the public school curriculum as a matter of justice and fairness. However, their case required some unusual exclusions. As a result of Justice Black's 1947 wall-of-separation metaphor, the defense could not use as witnesses the three acknowledged leaders of creationism, who had all written that creation-science is not any less scientific or any more religious than evolution and admitted outright that they supported creationism and opposed evolution for religious reasons. Indeed, because they used the term *religion* to explain their positions, their publications were cited in arguments against creation-science by the plaintiffs. Current constitutional interpretations of the First Amendment made the trial into a terminology battle turning on the difference between science and religion. During the trial, the term *religion* was so devalued in this key dichotomy that it was forbidden to appear. The Judge concurred, and the law was overturned.

How did important witnesses and trial watchers argue their definitions of science and religion, and why? This particular episode in the creation/evolution controversy involved not only a carefully scrutinized trial, but a whole new range of disciplinary knowledge brought to bear on a public problem through many acts of definition and persuasion. A study of some of these acts suggests that American culture has come a long way from Scopes yet not far enough to overcome its current opposition to rhetoric and its support of positivist philosophy. Indeed, society now uses academic disciplines to provide correct definitions of terms and to argue about worldviews and political philosophies.

HISTORY: GEORGE MARSDEN

A first important witness to study for his treatment of the trial is George Marsden, an expert on religious fundamentalism in America and a professor of history at Calvin College at the time of the trial. (Marsden is currently professor of the history of American Christianity at the Divinity School at Duke University.) Marsden was preceded at the trial by a Methodist bishop from Little Rock and an Old Testament scholar from DePaul University, but he was the first witness to publish widely about creationism in the years after the trial. Marsden's contributions to this controversy show that he defines *science* and *religion* as a historian and a political liberal who sees the creationists as opponents of history and liberalism. It is no surprise that he disagrees with them on these issues and decided to use his influence to oppose them in the public sphere.

Marsden was invited to the trial largely because of his 1980 book *Fundamentalism and American Culture*, which has been taken by many as the definitive 1980s study of American fundamentalism up to the time of the Scopes Trial.[34] In this book, Marsden characterizes fundamentalism as a particular branch of Christian evangelicalism that has defined itself in opposition to a liberal interpretation of the Bible and an increasingly historicist interpretation of Christianity. Marsden concurs that the chief spokesman of 1920s fundamentalism was William Jennings Bryan, who

spoke out on all these issues and decided to focus his attack on evolution because it encapsulated for him the major evils overtaking the Christian
churches and the entire culture. In this study Marsden "views fundamentalism not as a temporary social aberration, but as a genuine religious movement or tendency with deep roots and intelligible beliefs."[35] He was invited
to the Arkansas trial to explain the history and beliefs of this religious
movement and to show that creation-science is indeed a veiled version of
fundamentalist religion. Marsden and other members of the religion team
helped the plaintiffs argue that creation-science is not science but religion,
and thus could be banned from the public schools under the Establishment
Clause.

During the trial, Marsden testified that creationism is a key belief of fundamentalist religion; he agreed that it was the same basic stance defended
by Bryan in Scopes I.[36] But the plaintiffs received an added rhetorical advantage from his testimony when Marsden quoted a key creationist, Henry
Morris, as follows: "This philosophy [of evolution] is really the foundation
of the very rebellion of Satan himself and of every evil system which he has
devised since that time to oppose the sovereignty and grace of God in this
universe."[37] At this late date in the twentieth century, the words *good, evil,
God*, and *Satan* rarely appear in U.S. district courts. When they do, they
make many listeners uncomfortable, suggesting as they do a two-valued, absolutist, "good versus evil" mindset which has been regularly associated in
our century with dangerous political leaders and religious fanatics. Marsden's testimony proved the link between fundamentalism and creation-
science, by showing that creationists use simple-minded oppositions to defend dangerous political ideas.

Marsden has made this same argument in several essays about creationism published since the trial.[38] A first essay (which has been revised for different audiences and published in at least three different places) is entitled,
"A Case of the Excluded Middle: Creation versus Evolution in America." It
argues that creationism is a complex phenomenon with interesting historical roots—thus the need for a historian to write about it. Beginning with the
title, the essay further argues that the creationists created this controversy
themselves by polarizing the issues and successfully persuading most of the
American public that a conflict exists between creation and evolution. Reflecting his interest in the history of creationism and his desire to explain
this position to the reader, Marsden's historical analysis traces many important elements in its rise, including the roles of millenarianism, biblical literalism, Baconian science, and Scottish commonsense philosophy. Near the
end, he concedes that the creationists do have one legitimate complaint:

evolution has become "an all-explanatory metaphor in modern culture,"[39] or what David Livingstone (whom Marsden cites) has called a "full-blown mythology."[40] He objects to efforts to make evolution the basis for an entire worldview, but he does not ultimately accept the creationists' position because he does not share their belief in biblical inerrancy nor their resulting sense that creationism and evolution necessarily conflict. He implicitly encourages the creationists to study their history and learn that many religious leaders throughout the past hundred years have been able to reconcile creation and evolution simply by giving up their belief in biblical inerrancy. Moreover, he suggests that by surrendering this one irrational belief, the creationists will be able to accept evolution as a "biological theory [that] is not *necessarily* connected with [an anti-God] world view."[41]

This essay also provides a tactical suggestion to the evolutionists. In the version published in the science journal *Nature*, Marsden argues that the evolutionists have brought this controversy upon themselves by treating evolution as an all-encompassing account of reality. The essay concludes, "Dogmatic proponents of evolutionary anti-supernaturalistic mythology have been inviting responses in kind."[42] Marsden's solution to this controversy is to distinguish between evolution as science and evolution as mythology: we all must accept the former, but we are free to reject the latter.

In a second essay, entitled "Understanding Fundamentalist Views of Science," Marsden further clarifies his own conception of science. He argues that the creationists refuse to come into the twentieth century on this matter, clinging instead to a Baconian notion of science that has been radically revised (primarily by Darwin) since it first appeared in the seventeenth century. Marsden attacks the creationists because they distinguish between a *theory* as an unproven idea and a *fact* as an idea supported by objective proof. He argues that creationists need to update their definition of a *scientific theory*; evolution is clearly a scientific theory supported by facts. After historically analyzing changing conceptions of science, Marsden concludes with the same point that he made in the previous essay: that evolution must be accepted as science but need not be accepted as an anti-God worldview. He writes: "Once [it] is recognized that the scientific evidence is not going to settle the religious issue[,] useful discussions between those who believe in divine creation and those who do not may ensue."[43] His point is that both creationists and evolutionists need to distinguish the key terms *science* and *religion* so that they can begin to talk usefully again.

A final essay by Marsden applies his historical and definitional thesis to the American political system as a whole. Entitled "Secularism and the Public Square," this essay traces the rise of secular power in the United

States. Knowing that religious people usually react to this increasing secularization with concern, he reassures them that there is no need to worry: "As long as a progressive-liberal ethical system of non-judgmental openness and tolerance prevails, secularism seems no great threat."[44] In arguing for tolerance, a central value in liberal political philosophy, Marsden indicates that as long as American society continues to disempower the intolerant all will be well. He symbolizes this position both in his title and in an image drawn from his hometown in Pennsylvania, the state best known for its religious diversity. The courthouse in his town sits on the public square, while the various churches sit on lots separated from the public square. He proposes that "a worthwhile working principle" is "the *distancing* of religion from the public square."[45]

This essay effectively highlights a primary disagreement between Marsden and the creationists. In arguing for tolerance toward evolution as a scientific theory entitled to a place in the public square, he suggests that they give up their central belief in a literal Bible. This belief simply cannot coexist with a belief in tolerance; as Walter Lippmann wrote in his commentary on the Scopes Trial, "If you are open-minded about revelation you simply don't believe in it."[46] Creationists feel that they and their children should not have to tolerate the theory of evolution; it seems to them to contradict their belief in the revelation of a literal Bible. Thus, accepting this one form of tolerance would rob them of a cherished belief. The key word *tolerance* loosely means "to put up with something you do not like." Marsden is willing to put up with the creationists' religious beliefs so long as these beliefs remain in the private sphere. However, if the creationists attempt to introduce their beliefs into public schools, he implies, these beliefs must cease to be tolerated. In his commitment to the ideal of tolerance, Marsden accepts that the one thing not to be tolerated is intolerance itself, just as the one thing not to be countenanced in a political system committed to rationality is the idea that a belief in reason is not itself strictly rational. The creationists do not ultimately commit themselves, as Marsden does, to the key words *tolerant* and *rational*. They are committed to their own religious beliefs whether or not they can be defended as rational within other conceptions of reason and are tolerant according to liberal ideals. The question becomes whose terms and definitions ought to prevail. Liberals like Marsden conclude that the creationists are irrational and intolerant. Creationists conclude that the liberal ideals of reason and tolerance are being used against them in a manner that is irrational and intolerant and that, finally, is trying to outlaw their beliefs, which are very reasonable, and to exclude from the public schools their scientific account of humanity's origins.

As a religious historian, Marsden appears to admire liberalism, not in spite of, but because of its categorical decision to keep religion out of politics. He, too, wants religion to remain private.[47] Why would Marsden surrender the political power of religion, in this case to the liberal ideal of tolerance? He suggests one answer by mentioning European history: he knows that the political enactment of religious beliefs has often resulted in persecution, war, torture, and death. Perhaps religious scholars like Marsden want to keep the term *religion* from being used by religious zealots. They want to distinguish between legitimate religions and cults, to distance themselves from incidents like the Waco, Texas, massacre of almost one hundred Branch-Davidians in 1994, knowing that these incidents are cited as examples of the dangers of religion and are regularly used in arguments against religious scholarship, religious universities, and other things of importance to many scholars of religion. Marsden seems to want to define the term *religion* so that it excludes cults and other "eccentric belief systems" that embarrass him.[48] He is willing to give up a certain measure of political power for mainstream religions in favor of a liberal political philosophy that disempowers these fringe groups. He gives an exemplary statement in his essay on secularism: "The fact is that religion and power is a very volatile mix. If one combines the ability to force one's will on others with the absolute religious conviction that one is right, the chances of tyranny, revolution, or civil warfare are vastly increased. Religious convictions can all too easily become like a sort of wild card that one puts together with whatever political bias one happens to hold."[49] Marsden finally fears the power of "religious convictions," "political biases," and other irrational beliefs. These beliefs are like "wild cards" in the risky poker game of politics. Marsden wants tentative, patient, tolerant beliefs; he has learned from Enlightenment rationalism that some beliefs are scary and may well try to overthrow the reign of reason.

As a result of this conception of irrational beliefs, Marsden sees creationists as people who are fundamentally irrational rather than as people for whom the term *reason* means a different thing. Marsden reveals in his writings about creationism commitments to reason, democracy, and tolerance that effectively exclude creationism. By testifying at the trial and writing about it afterwards, Marsden used his knowledge and influence to argue for his own values and beliefs and his own conceptions of science and religion. He did not want the conceptions of creationists to prevail.

Although no other historians gave testimony in the Arkansas trial (one prepared for the trial but did not testify because his contribution came to be seen as inessential by the plaintiff's lawyers),[50] a number of historians from many specialties have also written works about creationism (and reviews of each other's works about creationism).[51] Like Marsden's essays, these works also treat the trial from the viewpoint of history, focusing especially on such issues as reason and tolerance. After all, historians are taught as professionals to see the implications of any issue for their field, and to produce meaningful works within the disciplinary boundaries established within a liberal democracy and a free market economy that defines academic disciplines in terms of liberal ideals of reason and tolerance. In his definitions of key terms like *science* and *religion* and his claims that both terms are misused by creationists, Marsden reveals some of his own commitments as a religious historian and a political liberal.

SOCIOLOGY: DOROTHY NELKIN

The next witness called by the plaintiffs was Dorothy Nelkin, a professor of sociology at Cornell University. Unlike Marsden, who began to write about creationism only after testifying at the trial, Nelkin had become interested in the creation/evolution controversy many years earlier. She indicates in the preface to her first book, *Science Textbook Controversies and the Politics of Equal Time* (1977), that this book "began simply out of curiosity about the creationists as a group of people who dared to represent themselves as scientists while challenging the most sacred assumptions and norms of the scientific establishment."[52] Nelkin's study of creationism as a social phenomenon established her as an expert on the sociology of science and the history of biology textbooks. She was invited to the trial to testify from a sociological standpoint that creationism is not science and to explain what she found in her studies of creationist science texts.

Nelkin revised her first book only slightly after the trial and published the revision as *The Creation Controversy* in 1982.[53] Both books focus on two case studies of early creationism. The first study involved a battle about biology texts fought before the California State Board of Education by a group of creationist parents in the 1960s and 1970s.[54] Instead of using legislative means to get creationism in the schools as the Arkansas creationists did, the California parents decided in 1963 to get involved in textbook selection. They ultimately convinced the state board of education to do three things: stamp a notice in all biology texts that evolution is not a fact but only a theory, request important revisions toning down evolution in the texts recommended for adoption (Nelkin reports in a later essay that one text "reduced the discussion of the origin of life from 2023 to 322 words and the Darwinian view of nature from 2750 to 296 words"), and recommend that creation-science be formally included in the state science curriculum.[55] After ten years of success, the movement slowed down somewhat in 1974 when the board decided not to require that creationism be taught in science courses. Nelkin argues that the central issue of the entire controversy has been who should have the authority to determine science curricula: the community of scientists themselves or the will of the people, as democratically determined? She points out that scientists are often vehemently opposed to political control of science education, a position that clearly increases the power and authority of the scientists themselves. Throughout her analyses, Nelkin argues that power issues are central to this controversy.

The other case study involved a fifth- and sixth-grade social science course named Man: A Course of Study (MACOS), which was developed by cognitive psychologist Jerome Bruner and his associates with a $4.8 million grant from the National Science Foundation.[56] Like the biology textbooks produced under the Biological Sciences Curriculum Study grant, the MACOS course was another post-*Sputnik* effort to improve American education by applying the concept of evolution to the social sciences. Nelkin reports that the MACOS course compared human behavior to animal behavior in such sensitive areas as "religion, reproduction, aggression, and murder" (it included units on "the life of the salmon, the family life of herring gulls, and the social behavior of baboons"); it also introduced students to radically different cultures (primarily the Netsilik Indians of Canada, who practice senilicide and infanticide). Its professed goal was to teach these young students "that neither behavior nor beliefs have absolute value apart from social or environmental context."[57]

Many parents found this course an appalling example of the way in which evolution adversely affects public school curricula and undermines the religious beliefs of their children. Nelkin writes: "The course [was] not only built on evolutionary assumptions, but it [denied] the existence of absolute values, thus explicitly teaching just those controversial ideas that fundamentalists have long suspected were implicit in the teaching of evolution."[58] These parents let their opinions about the course be known. First published in 1970, the course had been introduced by 1975 into at least 1,700 schools in forty-seven states; by that time, an increasing number of outraged citizens had convinced Congress to investigate this federally funded program. Congress conducted an investigation and scuttled the program immediately by failing to appropriate funds for it as part of their grant to the National Science Foundation.[59]

Using these kinds of details, Nelkin argues that such struggles are always conflicts between social groups. Her background in sociology leads her to see groups as communities that share common values and beliefs, mostly without much concern for whose beliefs are true and whose are false. Not being a scientist or a religious scholar herself, she has no reason to defend one group or to protect its terminologies as carefully as Marsden does. This outsider status in the debate made her an especially valuable witness for the plaintiffs.

In her testimony, Nelkin reviewed the findings of her earlier creationist studies and suggested that creationists are using science to advance their goals in an unscientific way. They are primarily concerned about evolutionists challenging their conceptions of reality while teaching their children, and so they seek for the prestige of scientific evidence to support their a priori assumption that God created the world. She also contributed to the plaintiff's case by testifying that the creationists publish two versions of most of their biology textbooks: one with biblical references and one without. She knew of no other science texts that follow this practice. The plaintiffs were able to use her testimony as another powerful plank in their main argument that creation-science is not science, but rather, religion.

In her works on creationism published after the trial (her 1982 book and two essays revised from it), Nelkin further analyzes the creation/evolution controversy as a power struggle between social groups and answers two important questions: why does this conflict focus on the public schools? what does it reveal about contemporary attitudes toward science?[60] In response to the first question, she observes that the public education system in the United States is built on "the abiding faith that schools are a means to remedy social problems and to bring about social reform. Education is often viewed as an ideological instrument, a means of changing social perceptions, such as racial or sexual prejudices. The perceived importance of education as an instrument of reform is also the basis of a powerful conservative reaction among those who seek to protect traditional ideologies."[61] Thus, she sees the public schools as a battleground for competing ideologies. In terms of this controversy, the evolutionists attempt to use the public schools as a platform from which to attack the ideology of conservative Christians, an attack that invariably upsets conservative Christian parents. If these parents decide to take action, they respond by using the same schools as a platform to attack the ideology of the liberal evolutionists. In short, both sides accuse each other of the same thing: unfairly using the schools for purposes of ideology. This is a very interesting attack, given that many people are convinced that public schools ought to teach ideology.

At the point where muddy thinking about ideology begins to appear (the argument can be loosely paraphrased, as "We want an ideology taught, but we don't like that word—it's Marxist—and besides, we only want our teachers to teach the truth because otherwise they might teach ideology"), this controversy seems more and more a head-to-head struggle between two groups that agree on many things but not on the particular ideology they want taught, about which they diametrically disagree. Because they live within the same Enlightenment universe, which is devoted to the objective reality of truth, they use virtually identical arguments and strategies to oppose each other, usually even using the same terms. This battle (and many battles fought in a liberal society) becomes a contest between freedom and imprisonment, right and wrong, equality and inequality, and democracy and tyranny, in which the side associated with the positive term is always one's own, and the negative term is always linked to the position of one's opponent. In the case of the Arkansas trial, the battle opposed freedom and censorship, fairness and bias, balanced treatment and preferential treatment, equal time and unequal time. Both sides claimed the positive term for their own position.[62]

In response to the second question about contemporary attitudes toward science, Nelkin is similarly illuminating. She thinks that scientists are largely to blame for the predicaments they are increasingly facing: "When faced with external political pressures, scientists often take refuge in reasserting the neutral character of their work and the irrelevance of political, social, or religious considerations."[63] In short, they deny that they are advancing particular values and beliefs, taking advantage of what Nelkin sees as the "relative [protection of science] from public scrutiny as compared to other social institutions."[64]

When this denial is challenged, some scientists accuse the challengers of being opposed to science itself. Convinced that their findings are simple matters of fact rather than inevitable reflections of values and beliefs, they imply that they should not have to persuade the culture as a whole to trust science and to support its efforts to solve contemporary problems. In a review of Nelkin's 1982 book, George E. Webb writes:

Nelkin's chief contribution in this study remains her perceptive analysis of the resurgence of antiscience attitudes in the past three decades. Recognizing that this attitude may not simply by dismissed as anti-intellectualism or irrationality, Nelkin examines the motivation behind creationists' demands and finds several themes. The resistance to science exemplified by these partisans stems from the perception that science and technology threaten traditional values. Creationists distrust and resent the scientific experts who oversee science education, and they argue instead for pluralistic and egalitarian educational decisions.[65]

To the fear and resentment of science recognized by Webb, one need only add a deep and abiding belief that true science reveals facts and stamps them with the imprimatur of absolute authority, and one has captured the pre-Darwinian conception that marks the creationist attitude toward science.

In her conclusion to one essay Nelkin writes:

> The cognitive obscurity and social isolation of science has left the public dazed, resentful of professional expertise, and therefore receptive to reactionary influence. In today's social context, creationism appears to fill a void. It uses representations that are well adapted to the twentieth century; it claims scientific respectability while arguing that science is as value-laden as other explanations; and it offers intellectual plausibility as well as salvation, and the authority of science as well as the certainty of scripture. It reflects the prevailing political ideology, and its influence is likely to persist.[66]

Creationists want to borrow the respectability of the term *science* itself, to trade on the positive valences it continues to hold as the most rational method of a rationalist society, while at the same time attacking the values of those scientists who oppose a literal Bible. They want to call *authoritative* and *certain* not only their readings of the Bible, but also their readings of the solar system, the world of living things, and even the Grand Canyon itself.[67] They want to join scientists is using these words as authority figures within our current rationalist culture. Their use (or misuse) of the term *science* reveals not a plot against reason, but a use of the very term *reason* in a way that finally reflects the "prevailing political ideology" of our culture. To oversimplify, the ideology of both science and creationism is positivism joined with liberalism; it is committed to the idea of an objective truth and sees rhetoric and religion as its archenemies.

Nelkin herself seems occasionally to distrust rhetoric. This distrust can be glimpsed in the previous quotation in the tone of such phrases as "intellectual plausibility." She seems to wish that the creationists were unable to appear plausible; she thus joins those Western philosophers who have posited the existence of an obvious truth rather than admitting with the rhetoricians that truth is very difficult to recognize. Rhetoricians hold that, in this world at least, one must ultimately decide whose account of truth to believe.

A similar distrust of belief is evident in the following passage, which was taken from the concluding chapter of *The Creation Controversy* (a chapter entitled "Science and Personal Beliefs"): "The persistence of creationism reminds us that beliefs need no evidence; that, indeed, people are most reluctant to surrender their personal convictions to a scientific world view...."

Creationists, as we have seen, avoid, debunk, or disregard information that would repudiate their preconceptions, preferring to deny evidence rather than to discard their beliefs."[68] In this quotation Nelkin reveals her own disapproval of the creationists, suggesting that they hold their beliefs in spite of contrary evidence. Based on her training as a sociologist, she interprets *science* as a sociological term, but her own distrust of beliefs held without evidence has led her to side with the evolutionists against the creationists in the trial and to analyze both sides sociologically in her subsequent writing.[69]

THEOLOGY: LANGDON GILKEY

The next witness at the trial was Langdon Gilkey, Shailer Matthews Professor of Theology at the University of Chicago Divinity School. Gilkey has been the most widely disseminated commentator on the trial; his works have appeared in several essay collections and in various magazines addressed to significantly different audiences.[70] He was invited to testify as a result of at least two projects completed before the trial began. One was his book *Maker of Heaven and Earth* (1959), which has been taken as the definitive treatment of the Christian doctrine of creation.[71] The other, an influential essay entitled, "Evolution and the Doctrine of Creation," was itself a revision of Chapter 2 of his previous book.[72] These works established Gilkey as an expert on the religious doctrine of creation and on the philosophical relationship between science and religion.

At the trial, he testified that creation-science is indeed religion rather than science. First he explained that science depends on a methodology that automatically excludes God from the explanation of any natural event (the same point was made in Chapter 2 in relation to Darwin). In short, he argued that creation-science could never be science because it disobeys the first law of scientific method. In developing his other main point (that creation-science is inevitably religious), he explained that the notion of creation makes no sense unless it is a religious notion; in fact, the Arkansas act committed blasphemy because it posited a creator and then denied that this creator was necessarily God, thus repeating the heresy of the second-century Gnostics.[73] This testimony created a sensation; indeed, it worked as an indirect argument that creation-science was not only bad science, but also heretical religion. Thus, Gilkey added another brick to the wall erected by the plaintiffs between the disguised religion they claimed was being established in public schools in creation-science and the pure science of evolution itself.[74]

After his testimony, Gilkey was invited not only to write about the trial, but to lecture about it extensively. As a result of these lectures (which, he reports, he gave in "many colleges, universities, and laboratories")(vi), a colleague encouraged him to write a book. This book, entitled *Creationism on Trial: Evolution and God at Little Rock*, is one of the most interesting rhetorical products of this controversy. It uses a variety of persuasive tactics to distinguish between science and religion and to argue for Gilkey's commitments as a theologian.

The book is divided into two main parts: six chapters narrating Gilkey's involvement in the trial, and two long chapters analyzing the issues he encountered there. His entertaining narrative section ought to become the paradigm for a new academic genre—the trial expert witness detective story, which sounds like a work jointly composed by Perry Mason, Nicolò Machiavelli, and Plato in an effort to keep Galileo from being burned at the stake. It begins with an account of the first phone call Gilkey received from the ACLU attorney and then details the work they did together from several months before the trial until Gilkey had to leave after giving his testimony to return to his classes (and before he heard the testimony of any defense witnesses). It reveals a multitude of Gilkey's impressions: his sense that he was acting out a Federal Express Commercial when he got an overnight letter from the ACLU (3); his relief that the Arkansas attorney who cross-examined him was smarter than his broad shoulders and blond hair had suggested (82); his impressions of the dress, physique, and relative intelligence of the other witnesses; even the moments when gasps in the courtroom revealed significant shock. This interesting narrative creates a complex persona—one who admires academics, dislikes athletes, has a good sense of humor, and enjoys the suspense his testimony is able to create—all while defending evolution as science and attacking creationism as religion.

A few detailed examples will suggest how this narrative creates an effective persona. While describing his first meeting with Tony Siano (the ACLU lawyer who worked with him), Gilkey reveals to the reader that he decided immediately that for the sake of the trial he was willing to take out the earring he always wears in his right ear (as a token of his solidarity with the world's sailors) and to put on a tie in order to defend "our common cause" (7). However, he decides not to tell this to the lawyer; he wants Siano to form his own impressions of the meanings of Gilkey's earring and dress. The scene implies that academics do not have to wear suits and ties as lawyers do; in fact, they can even wear earrings. Nonetheless, they are willing to appear more conventional when they are needed to defend the truth, such as in courts of law where evolution is battling against creationism.

In another scene, Gilkey narrates his arrival at the Arkansas airport, where he and the other witnesses were "made [to] feel like visiting heroes, strong and capable warriors, tested by larger battles in the great world, and come briefly to their town to settle this local skirmish quickly and efficiently on their behalf" (77). Gilkey quickly counteracts the hubris of this comparison: "Although this heroic image hardly fitted my own nervous self-estimation at the moment, it felt very good indeed—and I suppose we soon enough got to believe in it, at least a little" (77–78). This scene suggests the shifts between confidence and fear that distinguish Gilkey's persona. It also illustrates his use of pointed oppositions and literary motifs to make the story more vivid, including a description of himself as a gallant warrior come from larger battles to help solve a "local skirmish." His comparisons figure the trial as a single incident in a global war, and himself as an Anglo-Saxon warrior, a Medieval crusader, or perhaps a Western marshal, come to conquer the creationist enemies.

As a final example of Gilkey's persona, consider the scene in which he recounts his testimony before the court. During his cross-examination, the defense attorney questioned the distinction Gilkey had posited between science as an explanation of "how's" and religion as the explanation of "why's," using weather as a concrete example:

> "I see that [Gilkey's explanation of the mechanical causes of rain]," said Rick, "but what is the *why* question you spoke to us about? Is there any such question, as distinct from the *how* question?"
>
> "There certainly is," I said, *being granted a sudden inspiration, I have no idea where from*, "and it represents a most important personal question about rain showers [besides how they were caused by cold fronts and other air patterns]. It is 'Why did it have to happen on this, my wedding day?'"
>
> Since fortunately the courtroom, *and even the judge*, burst into laughter at this example, the difficult line of questioning that might have followed (for example, "Why, Professor Gilkey, *did* it have to rain on your wedding day?") never got started. (122, italics added)

In this scene, Gilkey expresses delight at his ability to confound the questioner and to amuse the courtroom, the judge, and by implication the reader. Then he humorously deflates his delight by admitting that his witty response could have invited difficult questions that he might not have been able to answer so well. Gilkey's most interesting claim is that he was suddenly inspired, he knew not where from. This claim implies one of three possibilities: either God actually spoke to him so that the evolutionists could win (not an impossible option given his commitment to the justice of their cause); he is more intelligent than he himself realizes; or he does not want to appear as intelligent as he really is and thus ascribes his "sudden inspiration" to the grace of God. In any of the three cases, Gilkey clearly reveals his own commitment to the obvious truth of this position, for which he has decided to fight. His persona never reveals any uncertainty about the justness of his cause, even while it reveals continuing (and amusing) trepidation about his ability to defend it.

This narrative thus uses an alternately confident and nervous persona to persuade the reader to accept the evolutionists' position. However, its main rhetorical strategy is to describe the trial as a confrontation between true and false meanings for *science* and *religion*. Gilkey writes that despite countless revisions of his own basic testimony, its major argument always remained the same: that creationism was not science, but the view of a particular religion. In the analytical essays that complete the book, he extends the argument; he indicates that science must stop when it reaches the point of ultimate questions, and that religion must stop when it begins to reveal details about God's method rather than just asserting his existence as a whole. Both institutions have what he calls a "demonic" power to destroy unless they learn to temper each other (202–206). Gilkey thus depicts the trial as a battle about disciplinary and terminological purity in which he wants both disciplines to help and heal each other. Indeed, he warns the reader that if the two institutions are not kept separate, one or the other may cause mass destruction (212).[75]

In this narrative Gilkey also indicates that he set up what could possibly be the most important moment in the Arkansas trial: the cross-examination of the first defense witness, the theologian Norman Geisler. After Geisler had given his testimony, the plaintiffs' attorney asked him whether he believed in a personal devil. Geisler admitted that he did. On further questions, Geisler added that he believed this devil had been sending unidentified flying objects (UFOs) to the earth, and that he had learned about these UFOs from *Reader's Digest*. These admissions were greeted with bursts of laughter in the court; they set up the most famous headline of the trial about a theologian who believed in satanic UFOs. To many, Geisler's admissions made creationism seem obviously ridiculous.

This moment in the Arkansas trial reenacts the cross-examination of William Jennings Bryan in the Scopes trial: both reveal the power of ridicule in a culture devoted to reason. A rationalist culture implicitly holds that the beliefs or opinions of anyone who does not follow the laws of reason

need not be taken seriously. Indeed, if these beliefs or opinions are sufficiently irrational, they need not be treated with respect. Such unspoken assumptions suggest that people who make fools of themselves deserve
whatever ridicule they get. At this moment in the Arkansas trial, Geisler
was made to appear irrational enough that the social norms of politeness
were temporarily suspended; as a result, everyone laughed at him, and his
credibility as a witness was undermined. In some ways this was the moment
of victory in the Arkansas trial, just as Bryan's cross-examination was the
moment of victory according to many accounts of the Scopes Trial. The extended treatment of these moments by the media suggests that ridicule is
one of the most powerful rhetorical tools in a rationalist culture. If a rational
person can ridicule his opponents as irrational in such a culture, he wins the
argument.

As with other people publicly ridiculed by the press, Geisler stated that
he would never forget this moment. He indicates in his own book about the
trial that he wrote it in part to counteract his treatment by the media: they
virtually ignored his two-hour testimony in order to focus on these few sec-
onds of cross-examination and thus made his confession the best-known
headline of the trial. Although most observers would consider this admission a serious tactical error, if not a disturbing admission, on Geisler's part,
the incident proves again that certain beliefs are not tolerated by a liberal
political system, and specifically, that the word *devil* is currently prohibited
from a court of law. The incident suggests at least that academic ethos is created in the United States by reading scholarly and reliable books and that it
is lost by admitting that one reads the *Reader's Digest*. At most, it suggests
that no academic can admit a belief in Satan, UFOs, or supernatural phenomena without risking the loss of credibility as an intelligent and educated
person.

In his narrative of the trial, Gilkey explains how he himself set up this important incident. He reports that he was asked during his preparations for
the trial to read a copy of Geisler's deposition and to make suggestions on
what to attack. He describes his reaction when he discovered Geisler's pretrial admissions about Satan and UFOs:

> I hooted with delight, sprang to the phone, and when I got Tony on the line,
> said to him, "The Lord has delivered him into our hands! Go after that whole
> sequence on the UFO's, the Devil, and the *Reader's Digest* as if your life de-
> pended on it. When he states on the stand his dependence on that most un-
> impeachable of all sources of American folk wisdom, his status as an expert
> will have dissolved completely away!"

Needless to say, I was elated later when, unable to hear Geisler's testimony at the trial, I read in the *New York Times* of December 12: "Dr. Geisler acknowledged under cross-examination that he believes in unidentified flying objects as 'Satanic manifestations for the purposes of deception.' He said, amid courtroom laughter, that an article in the *Reader's Digest* had confirmed their existence." (76–77)

Gilkey reports that he "hooted with delight" to find this proof of irrationality and "was elated" to see that the entire world was laughing at Geisler. Indeed, the rationalist world did laugh, thus suggesting a commitment to rationality and a dislike of insupportable beliefs.

In the analytical essays included in this book (and the other essays he has revised from them), Gilkey more methodically discusses the relationship between religion and science that he only outlined at the trial. He argues that although many people seem to consider religion a temporary social aberration that will steadily diminish in power as scientific knowledge advances, he considers it a permanent human search for meaning in existence. He argues that science itself has lost much of the glory it attained in the early years of this century, when it was associated with the inexorable march of progress. In contrast to the widespread belief of that period in the inherent goodness of science, he points out the damage done by science in such places as Hitler's Germany, Imperial Japan, Joseph Stalin's Russia, and the Ayatollah Ruhollah Khomeini's Iran (168). His major argument is that religion and science provide complementary types of knowledge and that both must recognize their own limitations and respect the contributions of the other.

Such an argument replays the central notions of liberal tolerance examined in the discussion of Marsden. By calling on the court (and the culture as a whole) to distinguish correctly between religion and science, Gilkey reenacts the rejection of rhetoric evident throughout this controversy. Instead of positing the issue in terms of the difficulty of recognizing truth, he makes a distinction between truth and error on the basis of his training as a theologian. He gives the correct definitions and then calls on science and religion to respect each other, to "live and let live" and to be mutually tolerant.

In a passage reminiscent of George Marsden, Gilkey suggests his reasons for preferring this particular account of science and religion. He writes about the Arkansas Trial: "Some of us were there to defend *reasonable, self-critical*, and *liberal* religion against its *irrational, intolerant*, and *absolutist* varieties" (78, italics added). Each of these adjectives reveals a key value for Gilkey. He implies that the religion compatible with creationism is irrational, intolerant, and absolutist, and that by contrast the religion compatible with evolution is reasonable, self-critical, tolerant, and tentative. These particular distinctions require that he reject creationism altogether as the one irrational position that must not be tolerated after all. Gilkey indicates at the end of his narrative that the judge accepted the distinctions he had drawn and included them in "an amazing intellectual document" (155) that reached what Gilkey considers the correct decision. Gilkey's work thus reveals some of his own crucial commitments as an individual and a theologian; he chose to testify for the defense in order to aid the cause of evolution against the dangers of creationism. He wrote his post-trial essays to reveal to his readers which side was using correct definitions, and thus to defend reason, justice, and truth by differentiating science and religion.

PHILOSOPHY: MICHAEL RUSE

With the testimony of Gilkey, the plaintiffs concluded the team of expert witnesses representing religion as a discipline. The first witness in the science team was philosopher of science Michael Ruse, a professor of the history and philosophy of science at the University of Guelph in Ontario, Canada. In some ways, Ruse was the perfect hinge between the religious testimony just given and the scientific testimony to come. Ruse was chosen to testify because he had published several books on Darwin and had written a book attacking creation-science before the trial began (on which he was cross-examined).[76] The plaintiffs having demonstrated throughout the previous testimony (and especially in testimony given by Gilkey) that creation-science was unquestionably religion, and possibly even heresy, they were ready now to prove that it was not science. Accordingly, they had Ruse provide a list of criteria—tentativeness, falsifiability, predictive power, and methodological exclusion of the supernatural—on which a science is based and then explain why creation-science does not fulfill these criteria. Ruse provided the plaintiffs' correct definition of science so that the judge would be able to follow the details provided by the following scientists. After the trial Ruse wrote a personal narrative of his experience that provides a useful first glimpse of his reasons for opposing creation-science.

Ruse's narrative, which was expanded from a three-page essay entitled "A Philosopher at the Monkey Trial" into a thirty-page essay entitled "A Philosopher's Day in Court," begins with his flight south toward Little Rock to appear in the trial.[77] After pausing to give seven pages of background information about Darwin, the Arkansas act, and the case, Ruse then begins to discuss his own first encounter with an ACLU attorney and his subsequent preparation for the trial. By providing this background, Ruse creates a less literary and more forthright persona than Gilkey, suggesting that he wants to share an interesting personal experience rather than to write a mystery or a romance. He further strengthens his ethos by candidly admitting his own reasons for fighting the creationists; he writes that he believes creationism is "a real intellectual and moral evil" (322), defended by "sleazy" people who are plagued with "dishonest stupidity," and adds that he was honored by this chance to "stand in defense of the nobility of science" (336). By so candidly admitting his reasons for opposing creationism, Ruse eliminates the sense that he is defending what everyone accepts as the truth.

Indeed, Ruse reveals his commitments very clearly throughout the essay. His view of religion is suggested in his claims that the English gave up on a literal Bible long before Darwin published the *Origin*; that "in times of stress and unhappiness," fundamentalist religious beliefs appeal to Southerners, who want "simplistic doctrines for support and comfort" (314); that "God did not give us our reason just to have us hide our heads in the arid, comforting sands of Genesis" (a metaphor that subtly attacks Genesis) (327); and finally, that "it is those who deny evolution who are anti-God, not those who affirm it" (334). Creationists would probably not accept any of these statements, especially the claim that evolutionists are more pro-God than they are, but Ruse is not writing for the creationists. Indeed, throughout the essay he reveals his considerable disagreements with them and makes no arguments for tolerance.

Using the same directness, he reveals his impressions of the ACLU lawyers, the other witnesses, and other significant aspects of the trial. He is suspicious of the claim that the New York law firm is fighting this case out of altruism (*"pro bonum,"* as he writes, mistaking the Latin from *pro bono publico*, "for the public good"). After all, the firm "specializes, very successfully, in aiding the biggest of American corporations in their aims of swallowing up all competitors, or in aiding the just-less-than-biggest companies in avoiding being swallowed up by predators" (319). He finally decides that the firm must have accepted the case primarily to improve its public image, or more concretely, its recruiting potential at the best law schools, by "softening the picture of an enterprise concerned solely with money and power" (319). In similarly candid assessments, he calls Langdon Gilkey "a rather trendy and superarticulate theologian" (326) and says that, after a disappointingly brief cross-examination, "rather than talking of gaps in the record, [Stephen Jay Gould] was condemned forever to be one" (338). These candid impressions create a believable persona whose practicality and humanity strengthen his case against creationism.

Besides providing refreshing perceptions of others, he includes in the narrative several endearing details about himself. He admits he has never taken a biology class in his life; he took math and physics in his Canadian high school because biology was considered the class for the least intelligent students (320). He writes that on the day of his testimony he woke up at 3:00 A.M. and watched *The Lone Ranger* and *Sunrise Semester* on television because he was so nervous (328). He also admits that he went on for too long on the stand because "like all professors I talk too much" (324), and that he was so intimidated by the smart defense attorney that "by the time he had finished with me I was as limp as a rag" (324). All these personal touches make the essay enjoyable, even for a reader who disagrees with Ruse. His persona does considerable persuasive work, no matter how one feels about his definitions of key terms. Perhaps he gives the most revealing human touch of all at the end of the essay. Ruse had confessed earlier that when he got to his hotel in Arkansas and found a free hospitality room ("i.e., unlimited free liquor"), he got a bit too drunk and irritated the attorneys (326). However, after the scientists had finished their testimony and the judge seemed certain to decide for the evolutionists, Ruse and the other witnesses no longer needed to practice moderation. That night they all went out as a group "to eat in a restaurant—to talk, to drink, to play. Toward the end of the evening, someone started singing, and before long we were all joined in chorus. Inevitably we launched into hymns. My experience is that liberals almost always have a good church background and that under the influence this comes to the fore" (338). The essay concludes with an image of the witnesses singing "Amazing Grace" and laughing at its allusion to 10,000 years. With these final details Ruse confirms that he is a liberal who is not too religious. His persona is very effective in getting a reader to trust him and take his side in this controversy.

In the last two chapters of his 1982 book, *Darwinism Defended*, Ruse reveals less humorously the basic commitments behind his refutation of creationism. In this work Ruse attacks believers in Genesis as foolish and defenders of creation-science as both foolish and wicked. After giving some detailed recapitulations and responses to creationist arguments (although in a manner not universally admired by professional biologists),[78] Ruse gives detailed explanations of three sets of reasons (grouped under the subjects *religion*, *morality*, and *knowledge*) why creationism should not be allowed in the schools. In showing his definitions of the key terms in this controversy, his book's most useful section is the third one.

Under the heading of *knowledge*, Ruse argues that creationism cannot be taught in the public schools because it is too false to be presented to impressionable students and would consequently corrupt the educational process as a whole.[79] Convinced that scientists can teach "the basic fact of evolution, together with illustrations of some of [its] proposed natural causes" in language that is "value-free," he argues that the "thirst for knowledge," which led to the discovery of evolution, is also what has "made us human beings, in the best sense of the word," and that consequently, "scientific creationism . . . is a betrayal of ourselves as human beings" (326–327). In this passage Ruse posits the existence of a value-free language that corresponds directly to reality, and he extends the chain of linked dichotomies from science/religion, beyond truth/error, to human/inhuman. He suggests that creationists ignore the facts and evidence so completely that they are not only wrong and irrational, but actually inhuman.

Next Ruse adds that the schools cannot teach "every possible idea that people have held," but only "the best-sifted and most firmly grounded ideas that we have, together with the tools to move inquiry forward." Why is this "sifting process" necessary? "Without careful control of the content of the curriculum, one cannot inform and guide young minds" (328). This last sentence reveals a major reason why evolutionists want to exclude creationism from the schools: they are afraid it will "guide young minds" in the wrong direction, and thus counteract their own goal for the schools: to teach young students to be scientific—to doubt and test ideas on the basis of reason. Ruse is firmly committed to rationalist ideals. He opposes the creationists because they do not share these vital commitments.

In his conclusion, Ruse broadens his defense of evolution. He writes:

> It [creationism] is not simply mistaken; it is corrosive. Teaching scientific creationism will *stunt abilities in all areas* . . . . It is an act of bad faith even to present such ideas as a possible basis of belief. . . . If scientific creationism is taught as a viable alternative, there cannot fail to be a *deadening of the critical faculties*. . . . Hence my fight is not just a fight for one scientific theory. It is a fight for *all knowledge*. (329, italics added)

Ruse argues that the "corrosive" idea of creationism cannot be taught because it would "stunt abilities in all areas" and "deaden critical faculties." He sees the fight for evolution as a fight for "all knowledge." Ruse may finally broaden and overstate his argument in a way that undermines its credibility, but he clearly does distinguish between scientific knowledge and mere religious belief on the basis of his training as a philosopher of science. He wants students to develop critical faculties that test hypotheses for evidence against a backdrop of skepticism, and clearly holds a radically different worldview than the creationists.

Besides these two major works, Ruse has written other professional essays about his experience at the Arkansas trial. (The trial also provided an occasion for essays by other philosophers.[80]) Especially interesting is a lengthy debate he entered into with two other philosophers about whether he had defined science correctly at the trial and whether he had done harm to their discipline by testifying at all. In a series of published interchanges between Ruse, Larry Laudan, and Philip Quinn, Ruse has defended the criteria for science he presented to Judge Overton against several attacks. His two opponents have questioned his motives for testifying and his particular formulations of scientific criteria. In fact, Quinn has repeated the claim that philosophy deals with knowledge rather than politics, insisting that Ruse should not have involved philosophy at all in the unavoidable politics of a public trial. Some philosophers hold that even philosophers should not specify correct word meanings in a political forum.

These interchanges and several other anticreationist essays about the controversy have been compiled by Ruse in his collection *But Is It Science? The Philosophical Question in the Creation/Evolution Controversy*. This book contains nine essays by Ruse out of a total of twenty-eight—over one-third of the 400 pages total. Even the title of the book reveals that American culture attempts to resolve such controversies by answering positivist questions, and thus by positing true and obvious definitions. In effect, Ruse's book shows how philosophy works as a profession. One important work of philosophers is to argue about correct word meanings. Just as Ruse did his work in the trial by testifying that creation-science is not science, some philosophers see their work as proving that term X is not term Y and that the culture can call on them as experts on terminology.[81]

SCIENCE: STEPHEN JAY GOULD

After Ruse had given his testimony, several scientists testified that within their own disciplines creation-science is not science: Francisco Ayala, a geneticist at the University of California at Davis; G. Brent Dalrymple, a foremost geologist of the U.S. Geological Survey; and Harold Morowitz, a biophysicist at Yale University.[82] The last scientific witness, Stephen Jay Gould, professor of geology and paleontology at Harvard University, has become, in effect, the nation's foremost critic of creationism, having spoken out in many forums, including several public television programs. Gould reports that he first became interested in the theory of evolution at the age of five when his father took him to see a full skeleton of Tyrannosaurus rex at the Museum of Natural History in New York. After graduating with a Ph.D. from Columbia University (where he completed his dissertation on trilobites), Gould went on to become what a cover story in *Newsweek* (published three months after Gould appeared in the Arkansas Trial) calls "perhaps America's foremost writer and thinker on evolution."[83]

Since 1974 Gould has written a monthly column entitled "This View of Life" (a famous allusion to a sentence from Darwin's *Origin*) in *Natural History* magazine and has also collected his essays into several volumes, one of which won the National Book Award in 1980. Moreover, he has written many other scientific works, including a book coauthored with Niles Eldredge (the author of one of the anticreationist books cited at the beginning of this chapter) that introduced the "punctuated equilibrium" theory of evolution to the scientific community. According to this theory, which takes issue with Darwin's gradualism, the evolution of a new species progresses so rapidly that it leaves virtually no evidence in the fossil record; a state of relative species "equilibrium" is "punctuated" by periods of rapid evolution, usually due to changes in the environment. Because of superficial similarities between this account and the creationist's argument that the lack of transitional fossils proves God's acts of special creation, Gould has also become the evolutionist most often cited by creationists as proof that evolution is wrong. As a foremost expert on evolution and the unhappy source of multiple creationist citations, Gould was a natural choice as a witness who would powerfully conclude the plaintiff's scientific testimony.

In his testimony, Gould argued that creation-science is not science, that the geologic column cannot be explained by a Genesis flood, and that the creationists misuse his arguments whenever they cite him in an attempt to cast doubt on the theory of evolution itself. He asserted that evolution is a fact of nature and that evolutionists only disagree about its particular mechanisms. Coming as it did after heated rejections of the creationist accounts of genetics, geology, biophysics, and biochemistry, Gould's explanation of transitional fossils completed the plaintiff's case that creation science is not science at all.

Besides testifying at the trial, Gould has undertaken many other language acts to persuade the country to accept his conception of science and reject creationism.[84] In fact, his first essay on creationism, "Evolution as Fact and Theory," appeared in *Discover* magazine in May 1981, seven months before the trial began. This essay is a classic attempt to defend a terminological distinction that is crucial for the entire controversy: the difference between *fact* and *theory*. According to the creationists, facts are discrete and objectively knowable entities that cannot be disputed, whereas theories are guesses put together to account for facts, which themselves have not yet achieved the status of knowledge. Gould's thesis in this essay is that creationists do not understand the correct scientific meaning of the word *fact* and thus confuse all the issues in the controversy. He argues that according to a correct definition of *fact*, evolution has been established as a fact beyond dispute and that the creationists merely cloud the issues when they refuse to accept it and continue to say that it is only a theory. He thinks they are deliberately trying to evade this truth or else to conceal it: "Faced with these facts of evolution and the philosophical bankruptcy of their own position, creationists rely upon distortion and innuendo to buttress their rhetorical claim."[85] By now it should come as no surprise that the word *rhetoric* appears in a crucial position in this sentence. Gould argues that the creationists are aware of the emptiness of their own position and deliberately choose to avoid an honest search for truth and to rely instead on the hollow claims of rhetoric.

In essays published after the trial, Gould has continued to attack rhetoric while describing his involvement and drawing various conclusions from what he learned at the trial. His essay "Moon, Mann, and Otto" was begun while he was still in Little Rock, on the morning after he had finished his testimony. Although he writes in this essay that the trial was going well, it reminded him that trials like *Scopes* and *McLean* always represent a political tragedy: they allow "the battle for evolution to be won in the court of public opinion" and encourage textbook publishers, the most "cowardly and conservative arm of the [publishing] industry" to "[dilute] or [eliminate] evolution from all popular high school texts in the United States."[86] As an example of this dilution, Gould explains what he learned by comparing his own high school biology textbook, a 1950s edition of *Modern Biology* (by Truman J. Moon, P. B. Mann, and J. H. Otto), to the original 1921 edition (published by Moon alone).

According to Gould, the earlier edition was bold and direct in its treatment of evolution, but the later edition used "pussyfooting" chapters to dilute evolution into "a giveaway and an intellectual sham." As a result of this change, "millions of children [were] deprived of their chance to study one of the most exciting and influential ideas in science" (283–284). In contrast to Moon's dilution, Gould quotes from T. H. Huxley, whom Moon himself quoted in his conclusion when telling the reader to "sit down before fact as a little child" (285). Gould relates the story of a letter that Huxley wrote to his close friend, the Reverend Charles Kingsley, courageously refusing to console himself by turning to God after the death of his youngest son. Instead, Huxley "committed himself to science as the only sure guide to truth about matters of fact" and concluded that only "science and her methods gave me a resting-place independent of authority and tradition" (285). Gould similarly has committed himself to the key scientific values of skepticism, testing, and evidence, and to the courageous profession of one's beliefs about the truth regardless of their popularity. He writes about the teachers' testimony he had just heard that morning in Little Rock: "God bless the dedicated teachers of this world," those who refuse to bow to "imposed antirationalism in [their] classrooms (289). The essay thus links creationists to religion, authority, tradition, and antirationalism and invokes blessings on those who defy these forces and continue to teach what they consider to be the truth.

Gould concludes this essay with a final quotation from Huxley, again written to Kingsley: "Truth is better than much profit. I have searched over the grounds of my belief, and if wife and child and name and fame were all to be lost to me one after the other as the penalty, still I will not lie" (289). Gould is convinced that the creationists are lying. Rather than surrendering to their lies, he encourages the reader to "sit down as a little child" before the truth of evolution, to admit that such scientific truths are better than rhetorical flourishes or economic rewards.

In other essays, Gould has made similar attacks on creationism and its misuse of the term *science*. In an essay that appeared in the *Atlantic Monthly* five months after the trial, entitled "Creationism: Genesis and Geology," Gould quotes Paul Ellwanger, the original drafter of Act 590, to prove that one can identify as "just so much rhetoric" all the "[creationists'] appeals for 'equal time,' the American way of fairness, and presenting [both the theories] and letting the kids decide."[87] Echoing the themes of Ruse and Kitcher that students in public schools must be protected from creationist errors, Gould suggests that such terms as *equal time*, *fairness*, and *letting the kids decide* are not efforts to achieve the ideals they suggest, but rhetorical tactics that conceal other goals.

In this same essay Gould advances another argument that is repeated throughout this controversy: that to reject evolution is to do an injustice to God. Gould argues that evolutionists conceive of God as "a clockwinder," whereas creationists conceive of God as "a bungler who continually [perturbs] his own system with later corrections" (133). Gould ends the essay with a telling anecdote from his Arkansas trip; a plumber came into his hotel room as he was preparing to leave in order to determine if the toilet was leaking into the room below. The man gave a naturalistic account of the workings of water pipes, and then confessed that he believed in a literal creation and a Noachian flood. Gould argues that this man would be "a poor (and unemployed) plumber" if he applied his religious beliefs to his occupation as he did to his knowledge of the earth, and asks that we "approach the physical history of the earth" with the same "logical and mechanistic" account we would use to trace a leak in plumbing (133). By comparing the creationist account of the earth to a case of bad plumbing, Gould reveals again his commitment to scientific explanations of physical phenomena; he thinks that a supernatural account of the creation of the earth undermines any admirable notion of God.

Although Gould rejects creationism in part because he shares important commitments to rational inquiry with other academics, he also defends evolution because of its direct effect on worldview. In some of his early works, Gould argues that evolution has been so important as a discovery, not in spite of, but because of, its religious implications. He contends that evolution can help people to develop a new worldview instead of traditional Christianity, a view that will help the world solve many of its most pressing problems.

Gould most clearly articulates this goal in the prologue to his first essay collection, *Ever Since Darwin* (1977). He writes that Darwinism set up a "challenge to a set of entrenched Western attitudes that we are not yet ready to abandon" by arguing "that evolution has no purpose," that it "has no direction," and that "matter is the ground of all existence; *mind, spirit,* and God as well, are just words that express the wondrous results of neuronal complexity."[88] Gould argues that evolution allows us to explain such religious notions as God and such classic dualisms as *body/spirit* (notions now labeled "just words") as the result of "neuronal complexity." What effect might such a new worldview have? Gould writes, "Yes, the world has been different since Darwin, but no less exciting, instructing, or uplifting; for if we cannot find purpose in nature, we will have to define it for ourselves" (13). By accepting evolution, Gould suggests that we will be able to define our own purpose rather than to think that God has a purpose for us, to learn the lesson that he calls "a common theme [of these essays]—Darwin's evolutionary perspective as an antidote for our cosmic arrogance" (14), and especially to abandon what he sees as our hubris in thinking that nature exists for our sake rather than that we are a chance by-product of natural processes. Gould suggests in these statements that he sees evolution as a new "cosmic myth" or worldview in precisely the way described by Marsden and Gilkey. He does not attempt to define mutually supportive roles for science and religion but rather to replace a set of "entrenched Western attitudes" most often associated with Western religion with a new set of attitudes derived from the biological sciences.

He defends this commitment in the first two essays of this collection. In "Darwin's Delay," Gould explains that Darwin held back the publication of the *Origin of Species* for twenty years; he was afraid it would reveal that "God [could] not be anything more than an illusion invented by an illusion" (25). In the next essay Gould contrasts the tentative beliefs held by the young Darwin to the "ideological passion" of Darwin's companion aboard HMS *Beagle*, Captain Robert Fitzroy. Gould explains that Fitzroy was an ardent Tory and a Christian with a "dogmatic insistence on the argument from design" (33); this fatal combination of firm beliefs so unhinged Fitzroy's mind that at the 1860 meeting where "Huxley creamed Bishop 'Soapy Sam' Wilberforce," he "stalked about" while "holding a Bible above his head and shouting, 'The Book, the Book.'" The essay concludes: "Five years later, he shot himself" (33). In referring to religion and politics as "ideological passions," Gould uses the same root for his noun as the Greek *pathos*; this root also appears in *pathology* and is the Greek term for an orator's appeal to the emotions of his audience. Gould suggests that a firm commitment to a rationalist worldview would lessen the effect of ideological passions, and possibly even save the life of people like Fitzroy who cling too firmly to irrational beliefs.

In *Ever Since Darwin* Gould also explains what he means by calling evolution an antidote to "cosmic arrogance." In brief, he holds that a view of man as a special creation of God has encouraged him to mistreat the earth. Archibald MacLeish expresses the same idea in a piece entitled "Brothers in the Eternal Cold," which was written to commemorate John Glenn's successful Mercury orbit. After Glenn had become the first human being to see the earth from space, MacLeish argued that everyone could now envision the earth as a "tiny life raft" upon which all men must survive as brothers. He contrasts this ecological view to man's ancient view that he was himself the center of the universe, a position from which he felt he could "pillage and murder and plunder as he desired." Gould seems to share MacLeish's view, implying that many people have used the Genesis notion of "man's dominion" as an excuse for ravaging the earth, and that evolution suggests a new and more promising ecological worldview.[89]

Gould's position finally depends upon his own faith that science can improve life on earth through the use of reason. Throughout his work, Gould expresses admiration for science because of its rationality; he includes in his 1987 book *An Urchin in the Storm* a unit entitled, "In Praise of Reason."[90] In his work one can especially see the creation/evolution controversy as a battle between competing worldviews. In his more recent works he has situated science as a work within history and has cautioned scientists against misusing their cultural power. In a 1988 essay entitled "Genesis and Geology," Gould recounts a battle between the English political leader William Gladstone and T. H. Huxley about possible parallels between Genesis and earth history. After a rich account of this debate and the limitations of mythmaking on both sides, Gould concludes: "Genesis and geology happen not to correspond very well. But it wouldn't matter if they did—for we would only learn something about the limits of our storytelling, not even the whisper of a lesson about the nature and meaning of life or God."[91] The last sentence of the essay suggests that "we had better pay mighty close attention to both" rather than choosing between Genesis and geology (20). Similarly, in recent essays about Williams Jennings Bryan and Bishop James Ussher, the man who estimated the date and time of creation as quoted in *Inherit the Wind*, Gould has further investigated these two creationists and discovered that "Bryan had correctly identified the problem" with scientists who "misuse [their considerable cultural] power in furthering a personal prejudice or a social goal" and that "Ussher's chronology was a work within the generous and liberal traditional of humanistic scholarship, not a restrictive document to impose authority."[92]

But even within these revisions of some of his earlier statements, Gould still attacks creationism as a rhetorical deception. He writes in "Genesis and Geology": "Biblical literalism will never go away, so long as cash flows and unreason retains its popularity" (12), thus suggesting that he still perceives this particular belief of the creationists as irrational, and that "our legislative victory over 'creation-science' [through the Supreme Court decision striking down the Louisiana Act] ended an important chapter in American social history," thus suggesting that creationism is doomed (12). For Gould, a belief in biblical literalism could not be rational. Trained as a scientist in recent conceptions of rhetoric versus philosophy, he only understands creationist arguments as lies and continues his work of separating science from religion while attacking the fundamentalist world view.

After Gould's testimony, the plaintiffs turned to their educational team, demonstrating that many professional educators reject creationism as vehemently as many professional religious scholars and scientists. In brief, these witnesses testified that they could not obey Arkansas Act 590; it required them to teach creationism as science when they did not consider it science. They concurred with all the previous witnesses that creation-science was not science at all but veiled religion attempting to force itself into science classes. The purpose of their testimony was to convince the judge that a third group of academics also vehemently opposed this law: teachers, who would be most affected by it because they would have to obey it.[93]

A troubling and variously defined dichotomy between science and religion thus appeared in slightly different forms in the testimony of every witness called for the plaintiffs at the Arkansas trial. Although each person defined the dichotomy in slightly different ways, they all agreed on its importance. Dorothy Nelkin and Phillip Johnson have both written perceptive comments about this crucial relationship. Nelkin writes: "The battles between evolutionists and creationists continue—today's expression of the perennial warfare between science and religion. Their persistence suggests that the truce between science and religion, based on the assumption that they deal with separate domains, may be a convenient but unrealistic myth."[94] Johnson calls this myth an "official story" advanced by the leaders of American culture; they thereby attempt to persuade people that this conflict does not exist, that "reasonable persons need have no fear that scientific *knowledge* conflicts with religious *belief*."[95] Whereas religion scholars like Marsden and Gilkey attempt to resolve this conflict, scientists and philosophers of science like Gould, Ruse, Kitcher, and Futuyma all protect science against religion; they all want their readers to discount religious beliefs when they conflict with scientific knowledge. If some evolutionists use their theory to teach students that the Christian God does not exist, it is no wonder that creationists object to this theory as an ideological instrument. At the same time, if creationists use their account to prove that there is a Christian God, it is no wonder that evolutionists like Gould object to creationism in equally vehement terms. Perhaps the question of whether science and religion really conflict is less important than the question of whether people think they conflict.

To extend the point further, all scholarly explanations of why science and religion do not conflict prove by their very existence that science and religion do conflict, at least in the perceptions of many people. If they did not conflict, no one would try to explain the conflict away. Such differing interpretations of this conflict suggest that the conflict itself is a rhetorical construction rather than a natural fact. Whereas creationists and evolutionists acknowledge the conflict and use it to prove their own positions, people in other disciplines may ignore the conflict or attempt to explain it away. Whose position on the conflict is the correct one? It all depends on how one defines the terms *science* and *religion*—in what context, for which audience, and with what goals? In the Scopes Trial, Bryan and Darrow argued their own definitions. The plaintiffs in the Arkansas trial called expert witnesses from many fields to define these terms. Whose definition is correct? Which discipline captures the difference between science and religion? These questions are positivist. Their answers depend on how one resolves this particular Derridean dichotomy.

THE CREATIONISTS

After the plaintiffs had concluded their case, the Arkansas attorney general, Steve Clark, began his defense. He called as witnesses Norman Geisler, a theologian; Larry Parker, an education professor; Scott Morrow, a biochemist; Jim Townley, a high school chemistry teacher; Wayne Frair, a biologist; Margaret Helder, a biologist; Donald Chittick, a chemist; Ariel Roth, a biologist; Harold Coffin, a geologist; Chandra Wickramasinghe, an astronomer; and Robert Gentry, a physicist. Although I will report their testimonies later on and analyze representative writings about their experiences at the trial, I begin with the people who have most often made the creationist case: Henry Morris, Wendell Bird, and Duane Gish. Although these three all wrote about the trial after its conclusion, they were also a crucial part of the trial in spite of their conspicuous absence; indeed, they were erased from trial participation as witnesses for the defense because their published statements were used to support the plaintiff's case.

These three creationist leaders are significant in part because they have all used the term *religion* in direct support for their position. They have claimed that evolution is as religious as creationism and that they have worked to develop creation-science because of their religious beliefs. Besides their claims regarding the devalued half of the science/religion dichotomy, they differ from the expert witnesses for the plaintiffs in another important way. Rather than holding professorships at major universities, they have all been employed as staff members of the Institute of Creation Research (ICR) located in El Cajon, California: Morris as the director, Bird as the staff attorney, and Gish as the associate director. This institute has been a major interpretive community of the creationists, just as the various academic disciplines were important interpretive communities for the plaintiffs' witnesses, who represented these disciplines in their arguments and their definitions of terms. Besides providing employment, the institute has published creationist books under the aegis of Creation-Life Publishers (CLP), a research journal named the *Creation Research Society Quarterly*, and the ICR newsletter, *Impact*. This institute has become the primary meeting place for creation-scientists, their "world headquarters." Its existence and influence, independent of any major university, suggests that creation-science is an interpretive community without the trappings of an

academic discipline, and thus without much of the power. These three creationist leaders cannot create ethos through the prestige associated with a
professorship at Yale or a book published by the University of Chicago Press;
they have to create their ethos from scratch. This fact helps them with people who do not admire academics but hurts them in a culture where academics provide expert definitions at trials.

Henry Morris first attained prominence in the movement after co-
authoring *The Genesis Flood* in 1961. A Ph.D. in hydraulic engineering from
the University of Minnesota, Morris chaired the Department of Civil Engineering at Virginia Polytechnic Institute for thirteen years before resigning
to found the ICR and lead the creationist movement (following publication
of his 1974 work, *Scientific Creationism*). By 1989 Morris had written at least
twenty-two books defending creationism and attacking evolution as the
most dangerous idea of the modern world, citing many of the same reasons
given by William Jennings Bryan. Morris succinctly reveals his view of this
controversy even in the title of his latest book: *The Long War against God:
The History and Impact of the Creation/Evolution Conflict.*[96] This and other
books show that Morris has become the historian of the creationist movement; indeed, in some ways he *is* its history.

Morris discusses the Arkansas trial itself in his earlier book, *History of
Modern Creationism* (1984), where he calls it "the media circus which the
humanists desired" and Judge Overton's decision a "polemic against creationists, fundamentalists, and the literal interpretation of Genesis." He also
points out that although he had no role in the proceedings, his works were
cited by many witnesses for the plaintiffs and by Judge Overton himself ten
times in the opinion.[97] Morris was indeed an invisible presence at the trial
who probably would have become visible if it had not been for the First
Amendment. Instead of testifying for the defense, his writings were used by
the plaintiffs to prove that creation-science is religious rather than scientific.

Wendell Bird, the creationist lawyer, first established his reputation in an
article he wrote for the *Yale Law Journal* as a student (an article that won the
Egger Prize and was written under the direction of Robert Bork, Yale Law
Professor and future unsuccessful nominee for the Supreme Court). This article, "Freedom of Religion and Science Instruction in Public Schools," argues (with 278 footnotes) that to exclude creation-science from public
schools is to establish evolution as a religion and thus to violate the First
Amendment. In a subsequent essay, "Freedom from Establishment and Unneutrality in Public School Instruction and Religious School Regulation,"
Bird complements his earlier discussion of what he calls "the proper con-

struction of the Free Exercise Clause" with a discussion of "the proper construction of the Establishment Clause."[98]

Besides interpreting the First Amendment in a manner consistent with creationism, Bird has also mapped out the legal strategy used by creationists in the 1980s. In an article entitled "Evolution in Public Schools and Creation in Students' Homes: What Creationists Can Do," Bird describes several possible creationist tactics.[99] To this article he appended the draft of a model resolution that eventually became Arkansas Act 590. Bird's articles use the same definitional strategies as the evolutionists but with a twist; he argues that both creationism and evolution are sciences *and* religions, and therefore that they should be equally included or excluded from the public schools.[100]

Bird offered to help Attorney General Steve Clark in the Arkansas trial, but Clark declined when Bird insisted that he be appointed an attorney of record in the case. This disagreement escalated into a minor scandal. Even as Clark was preparing his defense witnesses to testify, Bird was calling these witnesses at the last minute and advising them to leave town quietly because Clark was handling the case poorly and was certain to lose it. Bird warned them that their professional reputations would be damaged if they testified. As a result, one important witness—Dean Kenyon, a professor of biology at San Francisco State University—boarded a plane and disappeared the night before he was scheduled to appear in court.

Bird was later appointed by the Louisiana attorney general as chief defense attorney for the Louisiana Balanced-Treatment Act; he ultimately appealed this case all the way to the Supreme Court. After his stunning loss there, Bird published his 1,000-page trial brief, which focuses extensively on definitional issues involved in both legal cases.[101] In this brief he describes at length the many definitional and logical errors that he feels were made by the Arkansas plaintiffs and by Judge Overton. Although Bird had a major impact on the Arkansas trial and its issues, like Morris, he was an erased presence at the trial; only his writings were included in the court proceedings. They were cited several times throughout the trial and twice in Judge Overton's decision.

Duane Gish is generally considered the most persuasive exponent of contemporary creationism. Gish received a Ph.D. in biochemistry from the University of Southern California and worked for eighteen years as a biochemical and biomedical researcher at Upjohn Laboratories and at several universities (including Cornell University and the University of California at Berkeley) before taking a position at the ICR. In contrast to Morris's ethos of apocalyptic struggle and Bird's ethos as a master of legal definitions,

Gish creates his ethos in his numerous works about this controversy by presenting himself as a cool, methodical scientist calmly debating the evidence before an impartial audience. Gish is perhaps best known for his 1978 book *Evolution: The Fossils Say No!* which has been taken by antievolutionists as the most authoritative statement of contemporary creationism.[102] This book and Morris's *Scientific Creationism* have been attacked point by point in many anticreationist works, including the six monographs cited at the beginning of this chapter. One evolutionist paid Gish's book the backhanded compliment of calling it the "least inarticulate" of any creationist work.[103] This book was also cited by several trial witnesses in the Arkansas trial and by Judge Overton (twice). It provides perhaps the best example of the creationists' definitions of key terms and reasons for wanting creationism taught in science classes in public schools. Its thesis is that creationism and evolution are equally religious but that creationism makes better sense of the scientific evidence than evolution does.

The book begins with a chapter entitled "Evolution—A Philosophy, Not a Science," which traces the major creationist argument and sets up later chapters as evidence for this argument. The title itself suggests Gish's most important strategy: carefully, if idiosyncratically, defining his key terms. In the first two paragraphs, Gish defines *evolution* as "the theory that all living things have arisen by a materialistic evolutionary process from a single source which itself arose by a similar process from a dead, inanimate world"; in contrast, the creation model "postulates that all basic animal and plant types (the created kinds) were brought into existence by acts of a supernatural Creator using special processes which are not operative today."[104] Having defined this central dichotomy, he relates both terms to many other dichotomies, most crucially to theory/fact and science/religion.

Gish develops the theory/fact distinction immediately in the third paragraph: "Most scientists accept evolution, not as a theory, but as an established fact" (12). Citing as proof quotations from two noted evolutionists and the treatment of evolution as a fact by most science textbooks, he then points out that evolutionists not only deny the label *fact* to special creation but even the label *theory*. Gish argues in reverse that such labels should not even apply to evolution itself: "Not only is there a wealth of scientific evidence for rejecting evolution as a fact, but evolution does not even qualify as a scientific theory according to a strict definition of the latter" (12). The next several pages provide this strict definition (from the *Oxford English Dictionary*), which hearkens back to simplistic Baconian conceptions of science and ignores the history of science as a term since Darwin. He then attempts to prove in detail why evolution does not measure up to it. In brief,

Gish argues that since the evolutionary account of the origin of life was not observed by anyone personally, it cannot be tested experimentally, and it cannot be falsified. It thus does not qualify as science according to his conception.

What about creation as a theory? Does it equally fail the same three tests? Gish writes:

> Creation is, of course, unproven and unprovable by the methods of experimental science. Neither can it qualify, according to the above criteria, as a scientific theory, since creation would have been unobservable and would as a theory be nonfalsifiable. In the scientific realm, creation is, therefore, as is evolution, a postulate which may serve as a model to explain and correlate the evidence related to origins. (21–22)

Gish holds that by the same criteria he has applied to evolution, creation also fails. Both ideas are not *theories* but *postulates* or *models* within which one can explain the evidence; the best theory explains the most evidence. He ends this paragraph: "In fact, to many well-informed scientists, creation seems to be far superior to the evolution model as an explanation for origins" (22).

If neither evolution nor creationism qualify as scientific theories, how can one account for the popular acceptance of evolution as a scientific theory? Gish gives two reasons. First, "it is often stated that there are no reputable scientists who do not accept the theory of evolution. This is just one more false argument used to win converts to the theory" (22). According to Gish, evolutionists make the acceptance of evolution the measure of the scientists themselves; they attack the ethos of a creationist by claiming that anyone who rejects evolution is not a good scientist. In response to this ethical attack, he gives two pages of evidence to the effect that many scientists do reject evolution and then reverses an example often used by evolutionists as evidence against creationism: the fact that spontaneous generation and Ptolemaic astronomy were wrong but were nevertheless defended for generations by a majority of thinkers. Just as Copernicus and Galileo were persecuted by this majority, he holds that the creationists may also be the victims of a mistaken evolutionist majority.

Then Gish turns to his second explanation: "The reason that most scientists accept the theory of evolution is that most scientists prefer to believe a materialistic, naturalistic explanation for the origin of all living things" (24). Building upon the theological implications implicit in his choice of adjectives (*materialistic* and *naturalistic*), he next cites T. H. Huxley's son, Sir Julian Huxley, that "Gods are peripheral phenomena produced by evolution," and suggests that Huxley wanted "to establish a humanistic religion based on evolution" in which humanism would become "a nontheistic religion, a way of life" (25). These quotations suggest that Gish sees evolution as a religion attempting to overthrow Christianity.

As additional proof of this theological conclusion, Gish also cites Harvard evolutionist George Gaylord Simpson that Christianity is a "higher superstition" (than the superstitions of primitive peoples) and that "man stands alone in the universe," responsible to "no one but himself." Based on these quotations from Huxley and Simpson, Gish concludes: "They [a large majority of the scientific community] have then combined this evolution theory with humanistic philosophy and have clothed the whole with the term *science*. The product, a nontheistic religion, with evolutionary philosophy as its creed under the guise of *science*, is being taught in most public schools, colleges, and universities of the United States. It has become our unofficial state-sanctioned religion" (26). Gish holds that evolution has triumphed by associating itself with the privileged term *science* and distancing itself from the devalued term *religion*. In the meantime it has served for many (but not all) evolutionists as the basis for a contemporary religion of its own.

Besides showing that famous evolutionists deny the existence of God and attempt to set up an alternate religion, Gish suggests another reason to consider evolution itself a religion. He writes: "It is apparent that acceptance of creation requires an important element of faith. Of course, belief in evolution also requires a vitally important element of faith" (27). After giving as an example the evolutionary argument that the human brain developed 12 billion cells with 120 trillion connections as a sole result of the environment acting on "the properties inherent in electrons, protons, and neutrons," Gish asserts: "To believe *this* obviously requires a tremendous exercise of faith. Evolutionary theory is indeed no less religious nor more scientific than creation" (27). Because both accounts require faith and make statements about God, Gish concludes that both accounts are religious.

After arguing that this controversy thus pits two scientific and religious accounts against each other, Gish suggests how one should decide between these accounts. His conclusion to the first chapter outlines the rest of his book:

> The question is, then, who has more evidence for his faith, the creationist or the evolutionist? The scientific case for special creation, as we will show in the following pages, is much stronger than the case for evolution. The more I study and the more I learn, the more I become convinced that evolution is a

false theory and that special creation offers a much more satisfactory interpretive framework for correlating and explaining the scientific evidence related to origins. (27)

For Gish the choice between these accounts depends on evidence. By gathering and correlating evidence, he holds that one can determine which
faith has more support in the natural world. On the basis of his own study of
the evidence for both accounts, Gish makes a final ethical appeal for the
reader to reach the same conclusion that he has reached. Not as a religious
believer but as a rational thinker he is convinced that creationism explains
the origin of life better than the theory of evolution.

The rest of the book compiles Gish's evidence. Its six more chapters give
extensive treatments of embryology, homology, the geologic column, the
fossil record for other animals, and the fossil evidence for human evolution.
He treats these complex issues in what has seemed to many scientists as obvious ignorance and distortion and to many nonexpert readers as clear and
cogent detail. Returning again in the final chapter to the terminological
distinctions he posited in the first chapter, he asserts: "The 'fact of evolution' is actually the *faith* of evolutionists in their particular world view"
(173). Quoting from the "convinced evolutionist" T. H. Huxley (a favorite
source for both sides in this controversy) that all "*a priori* arguments against
Theism" are "devoid of reasonable foundation," Gish concludes the book
with the following two paragraphs:

> The refusal of the establishment within scientific and educational circles to
> consider creation as an alternative to evolution is thus based above all on the
> insistence upon a purely atheistic, materialistic, and mechanistic explana-
> tion for origins to the exclusion of an explanation based on theism. Restrict-
> ing the teaching concerning origins to this one particular view thus
> constitutes indoctrination in a religious philosophy. Constitutional guaran-
> tees are violated and *true science* is shackled in dogma.
>
> After many years of intense study of this problem of origins from a scien-
> tific viewpoint, I am convinced that *the facts of science* declare special crea-
> tion to be the only *logical* explanation of origins. (174, italics in the original)

Gish's vocabulary in this conclusion suggests that he cannot conceive of
evolution as a logical explanation or of the rejection of creationism as anything but a deliberate refusal to face facts. His book suggests that he shares
Philip Kitcher's and Michael Ruse's devotion to "true science" and "obvious
facts," although his definition of science would predate theirs by at least a
century. Gish sees the legal decision to exclude creationism not as a dis-

agreement about what counts as *freedom*, *truth*, or *dogma*, but as an effort to "shackle" "true science" and to replace it with a "dogma" that succeeds by "indoctrination" rather than persuasion. From his own perspective, Gish's position seems eminently logical and reasonable.

From the perspective developed in this study, Gish's view depends on the same rejection of rhetoric and the same notions of universal reason and correct word meanings accepted by the evolutionists. Both Gish and the evolutionists claim to use the plain language of reason to defend the truth. They simply disagree on the content of that truth. From within their shared Enlightenment conceptions of reason and justice, they can only explain their fundamental disagreement about what counts as truth as an inability to see that truth or an effort to use rhetoric to obscure it. Ironically, the disagreement itself could not exist if the truth were obvious. In that case, the evolutionists and the creationists could not disagree about it. If words were mirrors for meanings, they could not mean opposite things to the two sides.

Besides this book, Gish has written many other defenses of creationism. These defenses have appeared in the *Humanist*, *Christianity Today*, two collections of anticreationist essays, *Discover*, and *Science Digest*.[105] They all trace the same argument and rely on similar explications of correct definitions, quotations from scientists, examples from history, and methods of creating a scientific ethos. Gish expands or contracts his argument in each essay to appeal to different audiences and to give more or less evidence for creationism. His contributions to *Discover* and *Science Digest* formed one side of a published debate: the first with Gould (over "Evolution as Fact and Theory") and the second with Isaac Asimov (who wrote an anticreationist essay entitled "The 'Threat' of Scientific Creationism" for the *New York Times Magazine* in June 1981, after the ACLU suit had been filed).[106] In these two essays, Gish also methodically defines his key terms, traces his main argument, and presents evidence to show that creationism is strictly logical and rational. Gish's very aura of rationality has infuriated some of his opponents; one said in exasperation that a person like Gish "can utter more nonsense in five minutes than can be refuted in five hours."[107] Although many scientists reject his arguments, many nonscientists do not consider them nonsense. The five books written by evolutionists to attack Gish's claims suggest that only an expert scientist willing to write a book can fully explain why they are nonsense after all.

Besides these published debates, Gish has traveled around the country debating scientists before university audiences.[108] He has also appeared in at least two conferences—the First International Conference on Creationism, held in Pittsburgh in 1986, and a Pacific Division meeting of the

American Association for the Advancement of Science held in 1984—where he has had similar opportunities to present his ideas and to defend them in public.[109] Of the three creationist leaders, Gish would have been the obvious choice to testify at the trial. He had presented his case for creationism hundreds of times for over a decade and had developed a reputation as a rational person able to explain his stances clearly and persuasively; indeed, many nonexperts have accepted creationism because Gish seemed so trustworthy even though they could not evaluate his arguments for soundness. However, instead of testifying for the defense at the Arkansas trial about the scientific evidence for creationism, Gish was quoted by the plaintiffs to prove that creationism was not scientific, and therefore did not belong in the schools. Gish attended the trial as a consultant for the defense even though he did not appear as a witness. He was present behind the bar in person but in front of the bar only in citation, thus testifying against the creationist case formally while advancing the creationist case informally.

Like Morris, Gish wrote a brief essay (only three pages long) about the Arkansas trial. Entitled "What Actually Occurred at the Trial," this essay was published in the ICR Newsletter *Impact*, no. 105 (with no date); it focuses on admissions made by the plaintiff's witnesses in cross-examinations and on the testimony of defense witnesses, especially Norman Geisler, whose testimony "destroyed the plaintiffs' case" according "to many" observers, and Chandra Wickramasinghe, who "chided evolutionists for their arrogance and intolerance of creationist views." In two moments of rare anger, Gish expresses amazement at one evolution teacher's testimony that creationism "would confuse students and so should be avoided!" and argues that the plaintiff's witnesses "had apparently been coached by the ACLU staff of lawyers to maintain that they knew of no scientific evidence to support creation and that creation science was altogether religious." These last two quotations reveal that Gish cannot conceive of a witness who sees no evidence for creationism and who thinks it is only a religious position. He feels they must be lying at the behest of the ACLU. In this essay Gish concludes that the case was lost because "Judge Overton (as well as most of the news media) completely ignored the scientific evidence presented by the defense witnesses while accepting without question evidence offered by the plaintiffs' witnesses. Many remarks made by Judge Overton during the trial revealed his bias against the creationist side." Throughout this essay, Gish reveals his implicit trust in evidence and impartiality. He ascribes the evolutionists' inability to see the evidence for creationism not to a different conception of the nature of evidence about the origins of life, but to a bias against creationism and a refusal to see the evidence that creationists present. Although Gish does not agree with the evolutionists about the particular ideas that can be labeled *facts* and *theories*, he does agree that science deals in facts and in theories that represent truth. Gish is perhaps the best example of the creationist attitude toward science that is characterized by Nelkin as a mixture of trust and distrust. Gish trusts science but distrusts evolutionary scientists. He is mystified that the participants in the Arkansas trial let what he sees as their bias against God overrule their ability to be fair and to see indisputable evidence of the truth of creationism.

The witnesses who did testify at the Arkansas trial basically developed the case that Gish has made. They argued for creation-science, attempted to prove anticreationist bias, and tried to show that professional biologists have worked to exclude creationism for unscientific reasons. In brief, Norman Geisler, the lead witness, attempted to convince the judge that both evolution and creation are equally scientific and religious (although his testimony was severely undermined by his admission about belief in satanic UFOs). Assuming that Geisler had proved that creationism was scientific, the other witnesses testified from their own fields that creationism better describes certain scientific phenomena than evolution and that therefore creationism ought to be included in the public school curriculum as a matter of justice and fairness. Of these witnesses, only Geisler has written works directly about the trial; the biologist Wayne Frair, the chemist Donald Chittick, and the geologist Harold Coffin each have written a book that defends creationism in general with arguments similar to those presented by Duane Gish.[110] (Because these books say nothing directly about the Arkansas trial, they will not be considered further here.)

The defense witnesses' case for creationism is best represented in Norman Geisler's *The Creator in the Courtroom* and in a book entitled *Scopes II: The Great Debate* by Louisiana State Senator Bill Keith, who sponsored the Louisiana Balanced-Treatment Law. Geisler's book is a self-proclaimed effort to counter what he considered the biased media presentation of the Arkansas trial. Except for a preface, an epilogue, and a brief chapter listing specific critiques of several aspects of the trial ("the media, the bill, the trial attorneys, the judge, and the ruling"), the book is a compilation and summary of relevant trial documents assembled so that readers may make up their own minds.[111] Accordingly, the book includes copies of the act, its legislative history, excerpts from the legal briefs, extensive records of all the witnesses' testimonies and cross-examinations, the judge's ruling, and an appendix containing Geisler's own writings about the trial. In an especially interesting feature, the book summarizes the testimony of each defense witness and then provides media reports about that testimony from several sources. This collection is intended to allow the reader to evaluate reports in the media for fairness and completeness. Geisler's detailed summaries suggest that other accounts of the trial—especially the accounts considered here—either completely neglect or radically abbreviate the creationist case presented at the trial. He thus indirectly argues that neither the press nor the witnesses for the plaintiffs took the creationist witnesses or their arguments seriously.

Several defense witnesses testified about the existence of such professional exclusion and neglect. Chandra Wickramasinge testified that he and his collaborator, Sir Fred Hoyle, "had not published their research in standard scientific journals because the editors of those journals generally are closed-minded to anything which questions Darwinian ideas on the origin and development of life. Instead they chose to publish it in book form so their critics and they would be free to have exchange of ideas on the matter" (152). They had found the journals stifling to professional debate and decided instead to publish books (a combined total of about fifty so far). In another story about the gate-keeping function of professional science journals, Robert Gentry, a physicist at the Oak Ridge National Laboratory in Tennessee, testified (with evidence from copies of personal letters) that his research on polonium halos (traces left in rocks by the radioactive decay of this rare element) had initially been "published quickly in the leading scientific journals." However, when "the implications became clear" that his findings posed problems for the fundamental assumptions of radioactive dating, and possibly even argued for a very young earth,

> the journals suddenly became closed to him, and it took repeated efforts, and threats to tell the press about the bias, for Gentry to be allowed to publish the papers in the journals again.
>
> He said the general reaction of the evolutionist community was to discount the research, even though they could show no errors in it. He gave several examples of geologists who responded to his research by simply saying it must be wrong because if it were right it would require them to rethink all their theories about the age of the earth and the formation of the earth's geology—a hard task! (155)

In a footnote near the end of the book, Geisler reports: "Robert Gentry has been informed since the trial that his contract at the Oak Ridge National Laboratory (Tenn.) will not be renewed" (241).

These accounts of scientific exclusion are supported by an incident reported in the October 22, 1990, issue of the *Wall Street Journal* (page B3B) and the October 24, 1990, issue of the *New York Times* (page A18). According to these reports, a man was denied a job writing a regular column for the *Scientific American* because he admitted in his final job interview that he believed in creationism.[112] The reporter explains that although the man had already written several successful essays for this column (which deals with amateur science experiments) and was otherwise eminently qualified, the editor feared that by hiring this creationist, "'*Scientific American* might inadvertently put an imprimatur on creation-science,' which would jeopardize its credibility with biologists." An emeritus professor of biology at UCLA who was interviewed for this story agreed with the editor's assessments, arguing that the creationist's "philosophy could enter everything [he] does as science." Creationism so clearly opposes the values of practicing scientists that some scientists protect these values by not hiring professed creationists.

Besides summarizing the testimony for both sides in the trial and Judge Overton's decision, Geisler includes in his preface and his epilogue a few of his own comments on the trial and its denial that evolution is a religion. At one point he comments on an irony he perceived in the case. Overton's ruling suggested to Geisler that the words *God* and *creator* can no longer appear in public schools, yet these same words appear in Judge Overton's own court each day in the opening invocation ("God save the United States and this honorable court") and in the key sentence of the Declaration of Independence ("all men are endowed by their creator with certain inalienable rights") (40). Geisler wonders whether this legal decision excluding the word *creator* from biology classes may also prevent reading the Declaration of Independence in government classes. A final irony is that Darwin added the word *creator* to the last sentence of his second edition of the *Origin of Species*; the phrase indicates that life was "originally breathed by the Creator into a few forms or into one." If the Arkansas decision banished "the creator from the courtroom," Geisler suggests that it ought to problematize reading the concluding sentence from Darwin's seminal work, which began the creation/evolution controversy 130 years before the Arkansas trial.

Geisler was so upset by the decision that his epilogue almost advocates another American Revolution (192). Citing several Supreme Court cases earlier in the book that recognized secular humanism as a religion, he claims that Judge Overton's ruling holds "that only humanist beliefs, including non-theism, evolution, and naturalism, can be taught in public school science classes" (40). His own opposition to humanism is so strong that he thinks a revolution may be necessary to restore the freedom of religion that this decision has revoked.

Geisler was also infuriated by a quotation about his testimony in the *Washington Post*, which said that the case for the defense started with "a spectacular fireworks display," a phrase to which Geisler objected in the lead for a news report (16). At times Geisler seems ready to set off some fireworks of his own—of the type described in the national anthem by Francis Scott Key. However, in a book published in 1987, *Origin Science: A Proposal for the Creation/Evolution Controversy*, Geisler attempts to avert such a revolution by proposing a study known as *origin science* that will build a positive case for creationism instead of just presenting the negative evidence against evolution.[113] Apparently ignorant of current conceptions of science or of the difficulties involved in founding a new one out of the blue, Geisler has clearly continued his commitment to creationism and its associated worldview in spite of the ridicule he received at the Arkansas trial.

The book by Bill Keith, *Scopes II: The Great Debate*, is primarily an account of the Louisiana Balanced-Treatment Law, but it includes three chapters summarizing and analyzing the Arkansas trial. This book argues from these two legal episodes and from Keith's own experiences that, not the creationists, but the evolutionists are attempting to establish their religion in the public schools.

Keith reports that for twenty-three years he was the city editor for the Shreveport afternoon newspaper. Increasingly disconcerted by what he considered the secular bias of the press, he first became interested in creationism as a result of an incident that happened one day to his twelve-year-old son. During a biology class, this boy was asked what he believed about the origin of humanity. When he answered that he believed in creation rather than evolution, he was ridiculed and harassed by the substitute teacher, who threatened to take him to the principal for punishment if he ever mentioned creation again. Keith writes that his wife was reduced to tears that evening in telling him this story, and that the same night he vowed to do something about this infringement of academic freedom if he ever won a political office (3–4). He decided soon afterward to run for the Louisiana State Senate. When he was elected, he made creationism his major cause.

In this book Keith defends creationism by using a strategy clearly suggested before one reads the first page. His main title is *Scopes II*, suggesting that this episode is a reenactment of the famous "monkey trial." He dedicates the book to his wife, "who stands close by my side in this great battle for academic freedom," the main term he uses in his defense of creationism. His epigraph is taken from Darrow's testimony at the Scopes Trial: "It is bigotry to teach only one theory of origins. . . . Can the human mind be limited by law in its inquiry after truth?"[114] This epigraph invokes Darrow as a person in favor of teaching both creationism and evolution. The preface itself follows up on these leads by briefly recounting the Scopes Trial and linking Darrow to creationism and Bryan to the bigoted evolutionists of the present day. Assuming that most of his readers have learned about the Scopes Trial from *Inherit the Wind*, a film in which "the three leading characters in Scopes I were enshrined in the minds and hearts of the American people," Keith explains that in the Louisiana Trial "the issue of academic freedom remains the same" (viii). Keith thus provides yet another representation of the Scopes Trial as an epic battle, only in this case as a battle that creationists rather than evolutionists must refight. In both versions—the evolutionist versions (described in Chapter 3) and the creationist version (developed here by Keith)—Darrow is the hero, Bryan the villain, and academic freedom the cause; but in the two versions, these three elements represent diametrically opposite positions. By reversing the positions associated with Darrow and Bryan, Keith argues that creationism deserves the same fair hearing previously denied to evolutionists so that the same ideal—academic freedom—may be preserved.

Keith's argument about academic freedom clearly depends on his assumption that both accounts of origins are equally deserving of rational consideration. Throughout the book he argues for the rationality of creationism. To prove that in spite of this evidence, creationism is not getting a fair and rational hearing, he also cites experiences like that of his son described above. For him, the issue of academic freedom requires that creationism appear in the public schools.

In an argument similar to Keith's, several defense witnesses at the Arkansas trial testified that they were afraid to teach creation-science because it might cost them their jobs or subject them to recrimination within the schools.[115] Such opposition to their ideas seems to creationists an infringement on their freedom of religion, speech, and thought. By structuring his book as an argument for academic freedom, Keith suggests that this term is very important to creationists as well as evolutionists. Both sides use this term to defend their own positions. In the Scopes Trial the term was primarily used by evolutionists. But when the creationists used the same key term to defend their position in the Arkansas trial, they contended that the evolutionists shifted the key terms of the debate, arguing instead that the crucial issue was the nature of science, and that creation-science was not scientific enough (and too dangerous) to deserve a fair hearing in the schools. Both positions depend on value judgments: Should creation-science be heard in the interests of fairness, or censored in the interests of knowledge? Is the key issue freedom of religion or the nature of science? In the midst of this value conflict, both sides assert their power to choose and define the key terms with or without being fair to the history of these terms and thus to silence the key terms of their opponents.

Keith attributes the loss in the Arkansas trial to several factors (these factors were also mentioned by Geisler). One is what he considers the bias of Judge Overton. Keith indicates that Overton was a Methodist whose minister was the first witness for the plaintiffs and the son of a high school biology teacher who attended the entire trial. In fact, quoting these possible conflicts of interest, an editorial in the *Arkansas Gazette* asked Overton to disqualify himself before the trial began, but Overton refused.[116]

According to Keith, another important factor in the Arkansas loss was the performance of Steve Clark, the attorney general and chief defense attorney. Only weeks before the trial began, Steve Clark told the press that he had qualms about the act he was preparing to defend. Seven days before the trial he let the ACLU, his opponent in the trial, auction off a luncheon with him as part of a fundraiser to support their own trial efforts.[117] Both Keith and Geisler cite evidence that Clark turned down offers of free help from many different sources and only spent one-tenth of the time or money preparing his defense that was spent by the ACLU for the plaintiffs: about $30,000 compared to $2–$3 million (the largest sum of money—all pro bono—that the ACLU had ever spent on a single case).[118] Clark was also accused by the judge and several journalists of inadequately cross-examining witnesses for the plaintiffs and inadequately preparing the defense witnesses for direct testimony or cross-examination; he was accused by the Reverends Jerry Falwell and Pat Robertson of conspiring with the ACLU.

Morris, Gish, Bird, and several other creationists have used such evidence to conclude that Clark and Overton were biased and that the trial was unfair. They thus imply that a different defense attorney and a different judge might have changed the outcome. In contrast, Geisler concludes that the case could not possibly have been won given the contemporary climate toward religion in the United States. I agree with Geisler. Based on the current conception of the meaning of the term *religion* in the First Amendment, their disregard for historical changes in the term *science*, and the notions of universal reason and liberal tolerance that undergird our political system, the creationists could not have won. They were arguing from within notions of reason and justice particular to their own interpretive community, notions that seemed fair and rational to themselves but irrational and unjust to the culture as a whole. Lacking the trappings of professional expertise and the accompanying power to provide key definitions, they were attempting to defend a position within a culture of professional expertise based on principles inimical to that position. The fact that their argument was not heard (indeed, the argument was not even made by the people who have spent a lifetime making it) does not necessarily prove the irrationality of creationists but rather that both sides hold different worldviews within which they have differing conceptions of *reason, science, religion* and most of the other key terms.

Although both sides in the trial assumed that these key terms had obvious meanings, both also clearly made extensive efforts to specify the difference between *science* and *religion* and to defend their definitions. The fact that the key words meant different things to both sides proves that word meanings are not obvious after all but rather are constructed through efforts at persuasion. The arguments advanced by both the creationists and the evolutionists in the Arkansas trial suggest that word meanings could not be obvious facts about reality, but are instead rhetorical tools used in constructing that reality. In this trial the key words were used in particular ways by these particular human communities to reflect their differing commitments to a literal Bible in some kind of fundamentalist polity and to an academic search for knowledge in a liberal democracy. These terms were not labels for realities that are self-evident to humanity as a whole.

THE DECISION

A final text to be analyzed from this trial is the decision of Judge William Overton, the Methodist son of a biology teacher appointed to a federal bench in 1979 by President Jimmy Carter.[119] From his study, not far from this bench, he completed the most important rhetorical act in this episode of the controversy. Overton's decision shows that the evolutionists' witnesses and arguments persuaded him that creation-science was not science but religion and therefore had to be banned from Arkansas schools under the First Amendment. In exasperation, Norman Geisler has suggested that Overton's opinion could have been a revision of the ACLU pretrial brief.[120] The ability of this brief to persuade the judge could possibly indicate bias, but it suggests more clearly that the ACLU knows how to argue a case pitting science against religion by using key definitions of terms provided by experts in a compelling way.

The opinion begins by telling its own brief story of the act, the participants in the suit, and the trial.[121] It indicates that the plaintiffs brought suit against the law on three distinct grounds: its violation of the Establishment Clause, its violation of "a right to academic freedom," and its impermissible vagueness. The rest of the opinion is organized as a response to these three grounds, which are all framed as terminological issues.

In regard to the first ground, Overton begins by writing: "There is no controversy over the legal standards by which the Establishment Clause portion of this case must be judged" (257). In what could be taken as the most revealing sentence of the opinion, Overton indicates that the legal profession has unanimously decided to resolve such issues through terminological distinctions, and that this particular battle turns on whether creation-science qualifies as religion. Quoting Justice Black's opinion in *Everson v. Board of Education*, Overton then introduces the metaphor of a wall of separation between church and state. His decision for this trial is an effort to find and enforce that wall.

Next, he turns to a crucial 1971 Supreme Court case, *Lemon v. Kurtzman*, (403 U.S. 602), which established a three-part test of a law under the Establishment Clause. If the law fails any one of these tests, it can be ruled unconstitutional. In determining the status of Act 590 in regard to the first test (that "the statute must have a secular legislative purpose"), Overton pauses to give a history of fundamentalism compiled from the testimony of Marsden and Nelkin and a history of scientific creationism compiled from documents subpoenaed by the plaintiffs (he cites Morris, Gish, Bird, and Paul Ellwanger, the South Carolina creationist who revised Bird's model resolution into the Arkansas bill). Relying on these sources, Overton concludes that the creationist position does involve religious beliefs, and therefore that the act is religious in its intent. He ends this section of the decision as follows: "It [the act] was simply and purely an effort to introduce the Biblical version of creation into the public school curricula. The only inference that can be drawn from these circumstances is that the Act was passed with the specific purpose by the General Assembly of advancing religion. The Act therefore fails the first prong of the three-pronged test" (1264).

Overton's terms *simple* and *pure* use the same metaphor that was mentioned by professionals throughout the trial: both the theologians and the scientists who testified for the plaintiffs wanted to keep science and religion pure and distinct. Accepting their sharp distinctions rather than the blurrings of these distinctions suggested by the creationists, Overton concludes that the act advanced religion rather than science, and thus fails the test of secular purpose.

The next section of the decision considers the second test: whether the act advances religion as its primary effect. In a series of arguments that begin with the language of the act itself and then turn to testimony from both sides, Overton proves that the law does advance fundamentalist religion and a literal belief in the Bible. Overton rejects Geisler's key distinction that the notion of a creator does not necessarily imply a deity, writing that this argument "has no evidentiary or rational support" and comparing it to the heresies mentioned by Gilkey (1265). Relying further on the testimony of Gilkey, he argues that evolution deals with how life evolved without making any claims about why. Turning finally to the question of the nature of science (and the testimony of Ruse), he indicates that "creation science . . . is simply not science" because it does not use accepted scientific methods and fails to meet the "essential characteristics" of science, "the more general descriptions of 'what scientists think' and 'what scientists do'" (1268). In short, the law cannot have a scientific effect because creation-science is not science.

Throughout this section, Overton shows that he was not persuaded by defense witnesses like Robert Gentry. (Later in the opinion he calls Gentry's discovery of polonium halos a "minor mystery that will eventually be explained" [1270]). In response to their testimony of scientific exclusion, Overton writes:

> Some of the State's witnesses suggested that the scientific community was "close-minded" on the subject of creationism and that explained the lack of acceptance of the creation science arguments. Yet no witness produced a scientific article for which publication had been refused. Perhaps some members of the scientific community are resistant to new ideas. It is, however, inconceivable that such a loose knit group of independent thinkers in all the varied fields of science could, or would, so effectively censor new scientific thought. (1268)

Overton writes that the possibility of scientific censorship is "inconceivable." The creationists testified that they had not only conceived of such censorship but had actually encountered it. But since they could not prove that their articles were rejected by professional journals, Overton concludes that such censorship does not exist. In a critique of Overton's decision published in the *Journal of Law and Religion*, David Caudill argues that Overton did not consider in this case any of the recent work on science as persuasion.[122] Indeed, Overton did not hear any testimony based on the work of Thomas Kuhn, Bruno Latour, Paul Feyerabend, or any of the other philosophers of science (discussed in Chapter 2), who have argued that it is not only conceivable, but inevitable, that scientists trained in one paradigm should reject the work of scientists trained in another. In contrast to Ruse, these philosophers of science have accounted for scientific knowledge as a product of interpretive communities. However, Overton was unaware of this work on interpretive communities of scientists. By asserting that he found the thought of censorship inconceivable, Overton meant that he did not believe it. He did not think that a "loose knit group of independent thinkers" could effectively censor creationism or any other position that might constitute "new scientific thought."

After thus weighing the testimony and evidence presented by the scientists in the trial, Overton gives his conclusion regarding the second constitutional test: "since creation-science is not science, the conclusion is inescapable that the *only* real effect of Act 590 is the advancement of religion" (1272, italics his). In deciding that the act fails the second test, Overton asserts again that creation-science does not match the correct definition of the crucial term. In a culture which holds that words have precise meanings, even the concrete question of the effects of this law was decided by appealing to correct definitions of terms. The correct definitions for this trial were specified by a philosopher of science, Michael Ruse. He used his expertise to define *science* and to silence the competing definitions given by creationists for this most crucial of all the key terms.

Having concluded on the basis of, not one, but two, tests that Arkansas Act 590 violates the Establishment Clause, Judge Overton had completed the major interpretive work of this case. He used the rest of his decision to resolve a few other issues more briefly, including the third test. According to this test, an act is unconstitutional if it creates "excessive and prohibited entanglement [of government] with religion." Overton concludes that this act would create such entanglement by requiring the state to monitor teachers and science textbooks, presumably by determining if the texts and classes presented both sides fairly, kept religious aspects of creationism out of the class, and otherwise assured that the words of the law were enforced. This third test also suggests that our culture enforces its distinctions by carefully policing terminologies. All three of the tests became questions of the correct meanings and applications of key terms.

In other brief discussions Overton argues that "*balanced* is a word subject to ordinary understanding" (an expert need not define it); that he will not specify the meaning of "academic freedom" in this particular case; and that many phrases in the act are not "unconstitutionally vague," but "all too clear." In what may be the most important issue brought up by the defense, Overton asserts at last that even if evolution were indeed a religion, the state would not thereby be required to start teaching creationism but to ban evolutionism as well. However, he indicates, such a possibility need not be considered: "It is clearly established in the case law, and perhaps also in common sense, that evolution is not a religion and that teaching evolution does not violate the Establishment Clause" (1274). On the point most crucial to the creationists—the question of whether evolution has qualities often associated with religions—Overton asserts that evolution is not a religion and cites "case law" and "common sense" as his proof.

In referring to "common sense," Overton used a notion that in its Latin form provides the origin for an important synonym for law, *jurisprudence*. The term *common sense*, which, according to a poststructuralist conception means something like "the agreements that bind together an interpretive community," in this case ironically implied that by a self-evident standard of universal reason, everyone agreed on what counted as religion, and that the nonreligious nature of evolution was only "common sense." Rather than asking himself whether evolution could be religious under any definition, Overton simply asserted that it could not, thus showing that even though the law itself is based on notions of contingency and prudence (notions central to rhetoric throughout Western history), the American legal system speaks in the name of universals (a central notion of positivist philosophy). Overton thus demonstrated again that our culture is built upon the philosophy of John Locke and other Enlightenment rationalists and has subsequently set up solid walls to separate key terms such as *religion* and *science* rather than seeing these terms as fluid and rhetorical.

The decision concludes with a brief discussion about public opinion in relation to the First Amendment and a quotation from Justice Felix Frankfurter. According to a 1987 survey, the vast majority of Americans agree with the creationists that both creationism and evolutionism ought to be taught in order to assure fairness in the public schools.[123] Aware of this public opinion, Overton writes: "No group, no matter how large or small, may use the organs of government, of which the public schools are the most conspicuous and influential, to foist its religious beliefs on others." Unconvinced by creationist arguments that the evolutionists are using the schools to advance their own religious beliefs, Overton illustrates again that he sees the meanings of certain words as obvious and indisputable, indeed the very word (*religion*) that creationists have conceived differently and disputed throughout this trial.

The quotation from Justice Frankfurter asserts that "complete separation between the state and religion is best for the state and best for religion" because "if nowhere else, in the relation between church and state, 'good fences make good neighbors.'" Its last four words allude to Robert Frost's ironic poem, "Mending Wall," in which two New England farmers meet regularly to repair the wall that separates their property in order to preserve their good relations. By citing this allusion, Overton suggests again the phenomenal efforts made throughout this controversy to use language to erect fences and to police disciplinary boundaries. The church-state fence erected in this case did not please both the evolutionists and the creationists, as Overton's reading of the Frost poem implies. The evolutionists were pleased by their ability to fence creationism out of the public schools, but the creationists felt that this fence excluded their religious beliefs while allowing scientists to attack those beliefs under the guise of science, all while receiving legal protection and educational support from the state of Arkansas itself. Their key belief in a literal Bible had been judged irrational. They had lost the fight to uphold their worldview, at least through the method of passing a state law requiring equal time for creationism.

Although Judge Overton's decision was the authoritative text in this episode of the creation/evolution controversy, it was not the last text he produced in relation to the Arkansas trial. Soon after the trial he was invited to lecture to a group of Pennsylvania appellate judges at Bucknell University. At Bucknell he also gave a press conference in which he explained that he had received several death threats after the trial and now had to travel with a bodyguard (in this case, a federal marshal), thus illustrating in his fear for his very life that the work of persuasion goes on through other means when recourse to language fails.[124]

In his interesting and entertaining speech, Overton reveals much more. For one thing, he says the trial taught him that a typical creationist is not a "moonshiner in Eastern Tennessee" or "some Bible thumping preacher in a church down a dusty road in south Arkansas," but a college-educated person who has roots in "the high technology soils of southern California" and who faces a "serious dilemma" when his children are taught ideas in tax-supported schools that seem to contradict his "religious beliefs."[125] He sympathized with this dilemma, but he saw that the Arkansas law did not qualify as science and thus could not be upheld.

He also uses several images that reveal his solidarity with the academics and his distance from the fundamentalists. First he says the case was so easy that "the plaintiffs . . . ran through the state like the British took the Argentines" (10), thus comparing the efforts of the evolutionists to the recent conquest of the Falkland Islands by Great Britain. This image suggests vividly Overton's sense of the difference between the cultural influence possessed by the "great power" evolution and the "small power" creation; it also hints that both sides in this controversy had political motivations. In another passage Overton compares the plaintiffs' to "an all-star team" and the trial to "an academic camp meeting" (he cites Harper's as the source of this last image) (10–11). Comparing the respected professionals who testified at the trial to a basketball team that can beat any other team, and their gathering to a religious revival at which academics testify to their values, these images also suggest that the battle between evolution and creationism was indeed a political contest, a religious contest—even an endurance contest, which pitted a very small and weak creationist contingency against an overwhelming evolutionist competitor.

Overton concludes the speech by making two jokes, which I reverse in order—one about the amazing efficiency of the U.S. Postal Service, which managed to deliver a letter to him that was addressed only to "The Monkey Judge, Little Rock," and the other about the fact that, "within four hours of the time I filed the opinion in *McLean*, the Mississippi Senate passed a creation-science bill. So much for the persuasiveness of my decision" (21). Indeed, Overton did not persuade anyone of the meaning of such key terms as *science* and *religion*; he decreed what these words meant in this particular case, and thus upheld the key values of one group rather than the other. Perhaps his decision (and this episode of the creation/evolution controversy as a whole) has suggested more about our culture through its failure to persuade the creationists than it could have suggested through its successes—more about our fundamental assumptions regarding language, professionalism, and power.

NOTES

1. Among the other plaintiffs were Arkansas ministers of the Catholic, Methodist, Episcopal, Baptist, and African Methodist Episcopal Churches; the American Jewish Congress; the Union of American Hebrew Congregations; the Arkansas Education Association; the National Association of Biology Teachers; and the National Coalition for Public Education and Religious Liberty.

2. Niles Eldredge, *The Monkey Business: A Scientist Looks at Creationism* (New York: Washington Square Press, 1982); Douglas J. Futuyma, *Science on Trial: The Case for Evolution* (New York: Pantheon, 1983); Philip Kitcher, *Abusing Science: The Case against Creationism* (Cambridge: MIT Press, 1982); Chris McGowan, *In the Beginning: A Scientist Shows Why the Creationists are Wrong* (Toronto: Macmillan of Canada, 1983); Norman D. Newell, *Creation and Evolution: Myth and Reality* (New York: Columbia University Press, 1982); and Michael Ruse, *Darwinism Defended: A Guide to the Evolution Controversies* (Reading, Mass.: Addison-Wesley, 1982).

3. "The Evolution-Creation Science Controversy," *College Board Review* no. 123 (special issue) (spring 1982); "Creationism and Evolution," *Zygon* 22.2 (June 1987); and *Science, Technology, and Human Values* 7.40 ("Tenth Anniversary Issue with a Special Section on Creationism, Science, and the Law" (summer 1982). A special issue entitled "Evolution vs. Creationism: The Schools as Battleground," appeared before the trial in the *Humanist* 37.1 (January/February 1977).

4. Letter to the Reader, *Creation/Evolution*, issue 1 (summer 1980): i. This journal was edited by Frederick Edwords, administrator of the American Humanist Association, a group that has led much of the nonscientific opposition to creationism. The journal continued through twenty issues, until spring 1987.

5. Roland Mushat Frye, ed., *Is God a Creationist? The Religious Case against Creation-Science* (New York: Charles Scribner's Sons, 1987); Laurie R. Godfrey, ed., *Scientists Confront Creationism* (New York: Norton, 1983); Robert W. Hanson, ed., *Science and Creation: Geological, Theological, and Educational Perspectives* (New York: Macmillan, 1986); Francis B. Harrold and Raymond A. Eve, eds., *Cult Archeology and Creationism: Understanding Pseudo-Scientific Beliefs about the Past* (Iowa City: University of Iowa Press, 1987); Marcel C. La Follette, ed., *Creationism, Science, and the Law: The Arkansas Case* (Cambridge: MIT Press, 1983); Ernan McMullin, ed., *Evolution and Creation* (Notre Dame, Ind: University of Notre Dame Press, 1985); Ashley Montagu, ed., *Science and Creationism* (New York: Oxford University Press, 1984); Michael Ruse, ed., *But Is It Science? The Philosophical Question in the Creation/Evolution Controversy* (Buffalo, N.Y.: Prometheus Press, 1988); David B. Wilson, ed., *Did the Devil Make Darwin Do It? Modern Perspectives on the Creation-Evolution Controversy* (Ames: Iowa State University Press, 1983); and J. Peter Zetterberg, ed., *Evolution versus Creationism: The Public Education Controversy* (Phoenix, Ariz.: Oryx Press, 1983).

6. Readings from this course are compiled in Wilson, ed., *Did the Devil Make Darwin Do It?* Wilson explains the series of events that led to this course, including several debates between prominent creationists and evolutionists on campus and a skirmish in which "a creationist student [was] thrown out of [a biology] class [and] creationist students tried to have a biology professor thrown out of the university" (vii).

7. Stan Weinberg, ed., *Reviews of Thirty-One Creationist Books* (Syosset, N.Y.: National Center for Science Education, 1984). This collection was compiled in response to repeated efforts by the Iowa legislature to pass a creation-science law. In the preface Weinberg explains that "on the whole the reviews are unfavorable" because creation-science "tends to be meretricious pseudoscience(i)". Few other book review collections deny the existence of their subject.

8. A thorough study of the reaction to creationism in the press, popular magazines, and television is Marcel C. La Follette, "Creationism in the News: Mass Media Coverage of the Arkansas Trial," in *Creationism, Science, and the Law*, ed. La Follette, 189–208.

9. Tom McIver, *Anti-Evolution: An Annotated Bibliography* (London: McFarland, 1988); and Ernie Lazar, *Creation/Evolution Bibliography/Directory* (Sacramento: Author, [1987]).

10. Ronald L. Ecker, *Dictionary of Science and Creationism* (Buffalo: N.Y.: Prometheus Press, 1990).

11. These personal narratives will be cited and discussed later in this chapter.

12. Edward J. Larson provides a full legal treatment of the Louisiana trial in his excellent study, *Trial and Error: The American Controversy over Creation and Evolution*, updated ed. (New York: Oxford University Press, 1989), 166–184.

13. These quotations are taken, respectively, from reviews of Kitcher by John Habgood, "Myths of Religion, Myths of Science," *Nature* 300 (1982): 118; Michael Ruse, "Critical Notice," *Philosophy of Science* 51 (1984): 354; Duane E. Jeffrey, "Dealing with Creationism," *Evolution* 37 (1983): 1099; and Luther Val Giddings, "Penetrating Muddied Waters: Creationism and Evolution," *Dialogue: A Journal of Mormon Thought* 19.1 (1986): 176.

14. Kitcher, *Abusing Science*, 1. Subsequent page references to this introduction appear in parentheses in the text. All italics are added.

15. In another review of this book, Kenneth Habgood, the bishop of Durham, England, accuses Kitcher of perpetuating what he sees as the misleading myth of the Huxley/Wilberforce debate ("Myths of Religion, Myths of Science"). Paradoxically, Habgood's review does not attempt to protect the contemporary creationists from the same mythic misinterpretation, but only to rehabilitate Wilberforce by arguing that Wilberforce was not "the nineteenth-century equivalent of today's creationists" (118). The living English bishop apparently felt that he must protect the dead English bishop from slander by distancing him from the American fundamentalists.

16. Kitcher's phrase, "an article of religious faith," uses the same strategy described in relation to Dudley Field Malone's speech at the Scopes Trial. Both writers agree that science and religion may differ but imply that in cases of disagreement, science must take precedence. They suggest that religious people can continue to hold their beliefs so long as they see them as articles of religious faith, but they must not mistake them for scientific knowledge.

17. Douglas J. Futuyma, *Science on Trial: The Case for Evolution* (New York: Pantheon, 1983). Subsequent page numbers appear in parentheses in the text.

18. Norman L. Geisler, *The Creator in the Courtroom* (Milford, Mich.: Mott Media, 1982), x. The entire book will be analyzed later in this chapter.

19. Ibid., vii.

20. Dorothy Nelkin, *The Creation Controversy: Science or Scripture in the Schools* (New York: Norton, 1982), 44–47.

21. Stephen Jay Gould, *Ever Since Darwin* (New York: Norton, 1977), 11.

22. Nelkin, *Creation Controversy*, 46.

23. Ibid., 47. In his 1982 opinion on the Arkansas trial, Judge Overton describes the situation at that time: "The success of the BSCS effort is shown by the fact that fifty percent of American school children currently use BSCS books directly and the curriculum is incorporated indirectly in virtually all biology texts" (*McLean v. Arkansas Board of Education* 529 F.Supp. 1259).

24. John C. Whitcomb and Henry M. Morris, *The Genesis Flood: The Biblical Record and Its Scientific Implications* (Philadelphia: Presbyterian and Reformed Publishing, 1961).

25. Ibid., 1.

26. Henry M. Morris, ed., *Scientific Creationism*, public school ed. (San Diego: CLP Publishers, 1974). Morris claims to be the editor only but gives no indications of who else is responsible for which portions of the work.

27. Donald E. Boles, "Religion in the Schools: A Historical and Legal Perspective," in *Did the Devil Make Darwin Do It?* ed. Wilson, 170–188.

28. Ibid., 178.

29. In their article, "The Establishment of the Religion of Secular Humanism and Its First Amendment Implications," *Texas Tech Law Review* 10 (1978): 1–66, John W. Whitehead and John Conlan agree with Larson. They argue (with massive citations) that "the First Amendment was not meant to prevent the establishment of Christianity as a religion, but to prevent one Christian denomination from dominating the others" (3).

30. Larson, *Trial and Error*, 93–94.

31. John F. Wilson, "Original Intent and the Church-State Problem," Lecture, Duke University Divinity School, April 6, 1989.

32. The primary exception is the odd case of Mormon polygamy, but that story must wait for a different occasion. In the 1870s, the separation of church and state apparently was also an important issue, but in a way that did not involve the Establishment Clause. Boles reports that President Ulysses S. Grant proposed and the House of Representatives approved a Constitutional Amendment specifying new limits for the freedom of religion ("Religion in the School," 179–180). This amendment and its rhetorical context deserve further study.

33. Larson points out that although the free exercise of religion was an issue in the Scopes Trial, the judge could not have even conceived of—let alone accepted—a claim by the defense that the Tennessee Law violated the Establishment Clause of the First Amendment (*Trial and Error*, 93). This argument was not available within the 1925 legal system because Justice Black had not yet given his 1947 interpretation.

34. George Marsden, *Fundamentalism and American Culture: The Shaping of Twentieth-Century Evangelicalism, 1870–1925* (New York: Oxford University Press, 1980).

35. Ibid., 6.

36. My sources for testimonies from the witnesses are a transcript for the first part of the trial that I obtained from John Ball, Judge Overton's law clerk and now a staff attorney for the U.S. Senate Committee on Small Business, and detailed synopses of the trial written by eyewitnesses: LeRoy L. Sullivan, "The Arkansas Landmark Court Challenge of Creation Science," *College Board Review*, no. 123 (Spring 1982): 12–17, 32–35; Geisler, *Creator in the Courtroom*; and Bill Keith, *Scopes II: The Great Debate* (Shreveport, LA: Huntington House, 1982).

37. George Marsden, in Transcript, *McLean v. Arkansas Board of Education*, (529 F. Supp. 1255 E.D Ark. 1982), 82.

38. George Marsden, "A Case of the Excluded Middle: Creation versus Evolution in America," in *Uncivil Religion: Interreligious Hostility in America*, ed. Robert N. Bellah and Frederick E. Greenspan (New York: Crossroads Press, 1987), 132–155; "Understanding Fundamentalist Views of Science," in *Science and Creationism*, ed. Montagu, 95–116; and "Secularism and the Public Square," in *The Best of "This World*," ed. Michael A. Scully (Lanham, Md.: University Press of America, 1986), 103–117. A shorter version of the first essay is "Creation versus Evolution: No Middle Way," *Nature* 305 (1983): 571–574.

39. Marsden, "Case of the Excluded Middle," 148.

40. David Livingstone, *Darwin's Forgotten Defenders: The Encounter between Evangelical Theology and Evolutionary Thought* (Grand Rapids, Mich.: William B. Eerdmans, 1987). Livingstone thoroughly discusses evolution as a myth on pages 179–184. Livingstone's book is another history intended as an antidote for the historical ignorance of both creationists and evolutionists. In the concluding paragraph, Livingstone announces the following purpose for his book: "If only to curtail the abuses of rhetoric, creationists and evolutionists alike need to be made more aware of Darwin's forgotten defenders" (189). Livingstone is not the first to write a book devoted to protecting our culture from rhetoric, and he will probably not be the last.

41. Ibid., 148.

42. Marsden, "Creation versus Evolution," 1574.

43. Marsden, "Understanding Fundamentalist Views of Science," 112.

44. Marsden, "Secularism and the Public Square," 114.

45. Ibid., 116.

46. Walter Lippmann, *American Inquisitors: A Commentary on Dayton and Chicago* (New York: Macmillan, 1928), 64.

47. After reading this account of Marsden, Roger Roberts, a friend in the Duke University Religion Department and a student of George Marsden, told me that Marsden is generally taken by religious historians as attempting to return white male Christians to power in American culture rather than to disempower religious people. Roberts had not read these texts by Marsden and suggested that perhaps in this case Marsden relies on an argument *against* religion as embodied in creationism rather than presenting his usual argument *for* religion as embodied in his own reformed tradition. Our differing perceptions of Marsden reveal again that the word *religion* has different meanings in different situations. Roberts assured me that Marsden is not against religion in general but only against the religion practiced by the creationists. Thus, the word does not mean quite the same thing in the two different contexts, even for the same author.

48. Marsden, "Case of the Excluded Middle," 150.

49. Marsden, "Secularism and the Public Square," 114.

50. See S. Charles Bolton, "The Historian as Expert Witness: Creationism in Arkansas," *Public Historian* 4 (1982): 59–67. Bolton's interesting account suggests

that a good professional can find a way to write an essay about this exciting trial without even being fully involved.

51. Among other historical works dealing with the trial are Larson, *Trial and Error*; Willard B. Gatewood, Jr., "From Scopes to Creation-Science: The Decline and Revival of the Evolution Controversy," *South Atlantic Quarterly* 83 (1984): 363–383; James R. Moore, "Interpreting the New Creationism," *Michigan Quarterly Review* 22 (1983): 321–334; and three essays by Ronald L. Numbers: "The Creationists," in *God and Nature: Historical Essays on the Encounter between Christianity and Science*, ed. David C. Lindberg and Ronald L. Numbers (Berkeley: University of California Press, 1986), 391–423; "Creationism in 20th-Century America," *Science* 218 (1982): 538–544; and "The Dilemma of Evangelical Scientists," in *Evangelicalism and Modern America*, ed. George M. Marsden (Grand Rapids, Mich.: William B. Eerdmans, 1984), 150–160. A major book about the controversy by Numbers is discussed in Chapter 5. Book reviews include Willard B. Gatewood, Jr., review of Larson's *Trial and Error*, *Georgia Historical Quarterly* 70 (1986): 371–373; and James R. Moore, "An Equivocal Heritage," review of Livingstone's *Darwin's Forgotten Defenders*, *Science* 240 (1988): 1049–1050. Larson's book was also reviewed by a legal historian and federal appeals judge: Vito J. Titone, "Only Fools and Dead People Never Change Their Opinions," *Buffalo Law Review* 26 (1987): 193–209. A final interesting note about this intradisciplinary support network is that Numbers directed Larson's dissertation.

52. Dorothy Nelkin, *Science Textbook Controversies and the Politics of Equal Time* (Cambridge: MIT Press, 1977), x.

53. I will focus my analysis on this later book because it reveals, through its minor revisions, what she learned at the Arkansas trial. She expanded the earlier book of nine chapters into a book of twelve chapters by adding a chapter on the creationists' political tactics, a chapter on the Arkansas trial itself (which will not be considered here, as it does almost nothing but summarize the trial), and another new chapter about censorship entitled, "Censorship by Surrender." See Dorothy Nelkin, *The Creation Controversy* (New York: Norton, 1982).

54. This case study is summarized from ibid., ch. 7, "Creation versus Evolution: The California Controversy."

55. Dorothy Nelkin, "Science, Rationality, and the Creation/Evolution Dispute," in *Science and Creation*, ed. Hanson, 37.

56. This case study is summarized from Nelkin, *Creation Controversy*, 47–51.

57. Ibid., 49–50.

58. Ibid., 49–51.

59. Interestingly, this action was led by an Arizona congressman named John Conlan, who was aided in his efforts by a freshman senator from North Carolina—Jesse Helms. Helms apparently learned from this episode the arguments and tactics that he later applied almost verbatim to his 1989 campaign against financial support from the National Endowment for the Arts of what he considered

pornographic art. For Helms the issues were virtually identical, and thus the same rhetorical strategies could be applied.

60. Nelkin, *Creation Controversy*; Nelkin, "Science, Rationality, and the Creation/Evolution Dispute"; and Nelkin, "From Dayton to Little Rock: Creationism Evolves," in *Creationism, Science, and the Law*, ed. La Follette, 74–85.

61. Nelkin, *Creation Controversy*, 51.

62. Nelkin suggests (in *Creation Controversy*, 174–175) that the 1980s creationist banner of *equal time* originally came from the U.S. Federal Communications Commission's Fairness Doctrine, which holds that the media must grant equal time to both sides in a controversial issue and equal free airtime to each candidate in a political campaign. In "From Dayton to Little Rock," Nelkin adds that when the Arkansas "Balanced Treatment Act" lost, the original author of the act, Paul Ellwanger, went to work on an "unbiased presentation" revision (82), on the assumption that even if a good American can take a stance against *balance*, no American would defend *bias*.

63. Nelkin, "Science," 37.

64. Ibid., 33.

65. George W. Webb, review of La Follette's *Creationism, Science, and the Law* and Nelkin's *The Creation Controversy*, *Isis* 75 (1984): 581.

66. Nelkin, "Science," 44.

67. For the creationist reading of the Grand Canyon, see Thomas McIver, "A Creationist Walk through the Grand Canyon," *Creation/Evolution*, issue 20 (spring 1987): 1–42. The evolutionist reading is suggested by Giddings: "If growth rings in bristlecone pines, corals, or simply the sight of the Grand Canyon coupled with a little humble reflection don't negate creationist claims of a young earth, these Green River varves [alternating light and dark layers of sedimentation] certainly should" ("Penetrating Muddied Water," (175). Should they? Despite the claims about obviousness made by both sides, the Grand Canyon itself seems like a text that is not easy to read.

68. Nelkin, *Creation Controversy*, 191.

69. Another excellent sociological account of the Arkansas trial is Thomas F. Gieryn, George M. Bevins, and Stephen C. Zahr, "Professionalization of American Scientists: Public Science in the Creation/Evolution Trials," *American Sociological Review* 50 (1985): 392–409. This widely cited article explains the plaintiffs' strategy as an effort to place "a boundary between science and creation-science" (401) in order to "establish a professional monopoly over the market for scientific knowledge" (405) and thus protect "the continued public patronage of American science" (404). The authors add that the central strategy of evolutionists is to "[suggest] that the inclusion of creation-science in public schools threatened the effectiveness of science education by teaching error as truth" (406). It would be hard to improve upon the analysis made by this article. Some work by other sociologists relevant to the present study is discussed in Chapter 5.

70. Gilkey has published one book and at least four essays on the trial (some of which were revised and reprinted elsewhere). The book is *Creationism on Trial: Evolution and God at Little Rock* (Minneapolis: Winston Press, 1985). The essays are "Religion and Science in an Advanced Scientific Culture," *Zygon* 22 (1987): 165–178; "The Creationism Issue: A Theologian's View," in *Science and Creation*, Hanson, 174–188; "Creationism: The Roots of the Conflict," first published in *Science, Technology, and Human Values* 7 (1982): 67–71 and later revised for publication in *Is God a Creationist?* ed. Frye, 56–67, and in *Christianity and Crisis* 42.1 (1982): 108–115. In addition, his essay from *Science, Technology, and Human Values* was reprinted after another revision as "The Creationist Controversy: The Interrelationship of Inquiry and Belief" in La Follette, *Creationism, Science, and the Law*, ed. La Follette, 129–137.

71. Langdon Gilkey, *Maker of Heaven and Earth: A Study of the Christian Doctrine of Creation*, Christian Faith series (New York: Doubleday, 1959).

72. Gilkey, "Evolution and the Doctrine of Creation," in *Science and Religion*, ed. Ian Barbour (New York: Harper and Row, 1968), 159–181.

73. Gilkey, *Creationism on Trial*, 104. Further references to this work appear in parentheses in the text.

74. In another passage from *Creationism on Trial*, Gilkey hints at further interesting similarities between these two versions of science by figuring the court itself as a church. This passage indicates that the courtroom has reporters in "choir stalls"; "a large, bland, uniformed St. Peter at the doorway"; and an implied altar at the front (the bench), on which the black-robed judge sits (80). This passage lends support to the point made by James Moore that American culture metaphorically equates courts of law with courts of religion, and the verdicts rendered in courts of law to the verdicts that were once rendered by church officials, and will yet be rendered by God himself at the Final Judgment. Such spatial representations suggest that in our culture, scientists and judges stand in the place of clergymen and God.

75. Livingstone's nineteenth-century history has a similar comment about the importance of distinguishing between petroleum spirits and soda water, and of making sure that one does not drink one while thinking it is the other (*Darwin's Forgotten Defenders*, 184). Both Gilkey and Livingstone thus figure their own particular distinctions between science and religion to be vital to survival.

76. Ruse's most influential book has been *The Darwinian Revolution: Science Red in Tooth and Claw* (Chicago: University of Chicago Press, 1979); it was cited in Chapter 2. His attack on creationism forms the last two chapters of *Darwinism Defended: A Guide to the Evolution Controversies* (Reading, Mass.: Addison-Wesley, 1982).

77. "A Philosopher at the Monkey Trial," the brief essay, appeared in *New Scientist* 93 (1982): 317–319. It was written before the verdict was announced. The second essay, "A Philosopher's Day in Court," first appeared in *Science and*

Creationism, ed. Montagu, 311–342, and has been reprinted several times. Subsequent page references in parentheses refer to the longer essay.

78. Giddings concludes his review of Ruse as follows: "Here and there Ruse does provide entertaining bits of rhetoric in the 'call a spade a bloody shovel' vein, but his style would be [best] suited to the pages of the *National Enquirer*" ("Penetrating Muddied Waters," 173).

79. In a chapter entitled "Exploiting Tolerance" from *Abusing Science*, Kitcher reaches the same conclusion: that creationism must be kept out of public schools because it would confuse the students in their search for truth. In an excellent manuscript responding to this program entitled "Liberals and Creationists" (1989), Phillip Johnson, a law professor at the University of California at Berkeley, illustrates in vivid terms why this program seems riddled with inconsistencies for the creationists, who see it as yet another rhetorical strategy being used to censor their position. Johnson concludes that the evolutionists are in a very precarious position logically when they censor creationism from the public schools.

80. For example, Mary Midgley, a senior lecturer in philosophy at the University of Newcastle on Tyne, has argued that evolution is itself a religion: "Evolution as a Religion: A Comparison of Prophecies," *Zygon* 22 (1987): 179–194; and *Evolution as a Religion: Strange Hopes and Stranger Fears* (London: Methuen, 1985). In addition, another issue of *Zygon* was devoted to an essay by Martin Eger, a professor of physics and philosophy at the City University of New York (and several responses to this influential piece): "A Tale of Two Controversies: Dissonance in the Theory and Practice of Rationality," *Zygon* 23 (1988): 291–368. Eger uses the Arkansas trial to argue that the conception of rationality used in science differs from the conception used in moral discourse. His essay provides another example of MacIntyre's thesis that there is not one rationality, but multiple rationalities. It also suggests that philosophy sees itself as the study of the good as well as the true.

81. Besides the works analyzed above, Ruse has published another book on evolution entitled *Taking Darwin Seriously: A Naturalistic Approach to Philosophy* (Oxford, U.K.: Basil Blackwell, 1986). In the preface to this book, Ruse admits that the Arkansas trial convinced him that he knew almost nothing about "the foundations of morality." He writes: "In the months after the trial, because of the questions which I had been asked—questions which I had never truly asked myself—I grew to realize that at least my Creationist opponents had a sincerely articulated world picture. I had nothing." This preface is another example of Ruse's candor. He discovers what he does not know and sets out to learn it. This book is the result and the proof.

82. The sponsor of Act 590, Arkansas State Senator Jim Holsted, actually testified between two scientists, Ayala and Dalrymple, for reasons that are unclear to me. Holsted was asked about his purposes in introducing the bill and was discredited as another fundamentalist who attempted to change public policy with-

out first consulting the relevant experts, in this case professional educators. He was the only nonexpert who testified for the plaintiffs. See Sullivan, "Arkansas Landmark Court Challenge," 14–15.

83. Jerry Adler, "Enigmas of Evolution," *Newsweek*, 29 March 1982, 44.

84. These include three essays from *Hen's Teeth and Horse's Toes* (New York: Norton, 1983), entitled "Evolution as Fact and Theory" (253–262), "A Return to Dayton" (263–279) (analyzed in the previous chapter), and "Moon, Mann, and Otto" (280–290), which appeared in Part 5, "Science and Politics"; "Creationism: Genesis versus Geology," in *Science and Creationism*, ed. Montagu, 126–135; two essays from *Ever since Darwin* entitled "Darwin's Delay" (21–27) and "Darwin's Sea Change, or Five Years at the Captain's Table" (28–33); "Genesis and Geology," *Natural History* 97 (1988): 12–20; "Fall in the House of Ussher," in *Eight Little Piggies* (New York: Norton, 1993), 181–193; and "William Jennings Bryan's Last Campaign," in *Bully for Brontosaurus* (New York: Norton, 1991), 416–431.

85. Gould, "Evolution as Fact and Theory," 259.

86. Gould, "Moon, Mann, and Otto," 282. Subsequent page references appear in parentheses in the text.

87. Gould, "Creationism: Genesis versus Geology," 129. Subsequent page references appear in parentheses in the text.

88. Gould, *Ever Since Darwin*, 12–13. Subsequent page references appear in parentheses in the text. This same view that evolution denies design, purpose, and all realities except material ones is echoed by Futuyma, *Science on Trial*, 13.

89. It is doubtful that Genesis taught people to ravage the earth. Dorothy Nelkin points out that Christianity may have encouraged greater environmental sensitivity and protected the earth from ravishment by arguing (in contrast to other religions) that human are accountable to God in all things (*Creation Controversy*, 180 n. 14). Indeed, an argument can be made that the environmental ravishment caused by European colonialism and industrial capitalism began after the Christian vision of the "peaceable kingdom" had been powerfully undercut. The weakening of this Christian view resulted in part from Darwin's *Origin* itself; this evolutionary account was used for decades as a basis for social Darwinist policies before Gould or other scientists began to object to what they took as this misappropriation of Darwin.

90. An interesting contrast to Gould's faith in reason is *The Arrogance of Humanism* (New York: Oxford University Press, 1978) by the evolutionist David Ehrenfeld. This book paints a bleak picture of humanity's powers of reason, attacking especially the reasonableness of what Ehrenfeld takes as the central assumption of humanism: that we can solve all problems. He gives many examples of the irreversible errors made by scientists who claimed that they could resolve various problems but failed. He also argues that capitalism and communism are both economic systems developed from the same humanist assumptions and that by many criteria, humanism itself is the dominant religion of our time.

91. Gould, "Genesis and Geology," 20. Subsequent page references appear in parentheses in the text.

92. Gould, "Fall in the House of Ussher," 192; and "William Jennings Bryan's Last Campaign," 429–430.

93. The only educational witness who has published about the trial is William V. Mayer, head of the Biological Sciences Curriculum Study, and thus arguably a practicing evolutionary scientist rather than an educator. Mayer's essays are "Evolution: Yesterday, Today, Tomorrow," *Humanist* (37.1 (1977): 16–22, and "The Emperor's New Clothes—Sold Again," *Humanist* (37.6 [1977]: 52–53. Another essay has been published by an educator who was not a witness: Franklin Parker, "Behind the Creation-Evolution Controversy," *College Board Review*, no. 123 (spring 1982): 18–21. Parker is Benedum Professor of Education at West Virginia University. Both essays argue that creationism is not science, and therefore that it poses a threat to science teaching.

94. Nelkin, *Creation Controversy*, 194.

95. Phillip E. Johnson, "Evolution as Dogma: The Establishment of Naturalism," *First Things*, October 1990, 15–22. In a published response to this essay in the same issue (23–24), the biologist William Provine agrees that science and religion do conflict and that this conflict ought to be faced directly.

96. Henry M. Morris, *The Long War against God: The History and Impact of the Creation/Evolution Conflict* (Grand Rapids, Mich.: Baker Book House, 1989).

97. Henry M. Morris, *History of Modern Creationism* (San Diego: Master Book Publishers, 1984), 289–291.

98. Wendell R. Bird, "Freedom of Religion and Science Instruction in Public Schools," *Yale Law Journal* 87 (1978): 515–570, and Bird, "Freedom from Establishment and Unneutrality in Public School Instruction and Religious School Regulation," *Harvard Journal of Law and Public Policy* 2 (1979): 125-205. The quotations come from the latter article, pages 126–127.

99. Wendell R. Bird, "Evolution in Public Schools and Creation in Students' Homes: What Creationists Can Do, Parts I and II," *The Decade of Creation*, ed. Henry M. Morris and Donald H. Rohrer (San Diego: Creation-Life Publishers, 1981), 119–130.

100. Bird's legal writings have also inspired a response that uses definitional strategies against him: Delos B. McKown, "Creationism and the First Amendment," *Creation/Evolution*, issue 7 (winter 1982): 24–32.

101. Wendell R. Bird, *The Origin of Species Revisited*, 2 vols. (New York: Philosophical Library, 1994).

102. Duane T. Gish, *Evolution? The Fossils Say No!* public school ed. (San Diego: Creation-Life Publishers, 1978).

103. Giddings, "Penetrating Muddied Waters," 175.

104. Gish, *Evolution? The Fossils Say No!* 11. Subsequent page references appear in parentheses in the text.

105. Duane T. Gish, "The Scopes Trial in Reverse," *Humanist* 37.6 (1977): 50–51; "It Is Either 'In the Beginning, God'—or '. . . Hydrogen,' " *Christianity Today* 27.16 (1982): 28–33; "Creation, Evolution and Public Education," in *Evolution versus Creationism*, ed. Zetterberg, 177–191; "Creation, Evolution, and the Historical Evidence," in *But Is It Science?* ed. Ruse, 266–282; "A Reply to Gould," *Discover* 2.7 (1981): 6; and "The Genesis War," *Science Digest* 89.9 (1981): 82–87.

106. In regard to his *Discover* essay, Gish later wrote that he requested a chance to write an article of the same length as Gould's—four pages—but was told by the editor that the magazine would only publish a one-page letter to the editor. See *Evolution: The Challenge of the Fossil Record* (El Cajon, Calif: Creation-Life Publishers, 1985), 238. In contrast, the *Science Digest* debate printed internal reactions by each writer to the ongoing arguments of the other writer in a narrow column printed on the same page. "The 'Threat' of Scientific Creationism; Asimov's *New York Times* essay, which apparently set up this exchange, has been reprinted in *Science and Creationism*, ed. Montagu, 182–193.

107. Quoted in Giddings, "Penetrating the Muddied Waters," 176.

108. An entire book about the creationist debates is Marvin L. Lubenow, *"From Fish to Gish": The Exciting Drama of a Decade of Creation-Evolution Debates* (San Diego: CLP Publishers, 1983). This book describes many of Gish's 115 debates against evolutionists (including many of the evolutionists mentioned in this chapter: Michael Ruse, Chris McGowan, Ashley Montagu, Harold Morowitz, Frederick Edwords, Donald J. Weinshank, and William V. Mayer). The book has two appendices, which list every debate by place and by participant.

109. Creation Science Fellowship, *Proceedings of the First International Conference on Creationism*, 2 vols. (Pittsburgh: Creation Science Fellowship, 1986); and American Association for the Advancement of Science, *Evolutionists Confront Creationists* (San Francisco: Authors, 1984). At the start of his presentation to the second conference, Gish pointed out that the organizer only allowed two creationists but eight evolutionists to participate, even after extending the meeting an extra-half day to accommodate two more evolutionists. Gish concludes this aside: "I will proceed to take one of the two seats on the back of the bus reserved for the creationists at this meeting" (25–26). By not allotting creationists equal time, scientists thus reveal their own evaluations of this position: they do not consider it worthy of significant attention.

In addition to these conferences, a special session entitled "The Creationist Attack on Science" was sponsored by the American Society of Biological Chemists. No creationists were invited to this session, which was held April 19, 1982, at the Sixty-Sixty Annual Meeting of the Federation of American Societies for Experimental Biology (FASEB). See "The Creationist Attack on Science," *FASEB Proceedings* 42 (1983): 3022–3042.

110. Wayne Frair and Percival Davis, *The Case for Creation* (Chicago: Moody Press, 1983); Donald E. Chittick, *The Controversy: Roots of the Creation/Evolution*

Conflict (Portland, OR: Multnomah Press, 1984); and Harold G. Coffin, *Origin by Design* (Washington, D.C.: Review and Herald Publishing Association, 1983).

111. Geisler, *Creator in the Courtroom*, 13. Subsequent page references appear in parentheses in the text.

112. The March 1991 issue of *Harper's* has a partial transcript of one conversation between the editor and the creationist, which the creationist surreptitiously taped. ("Science's Litmus Test," 28–32). After the creationist accuses the editor of discrimination, the editor refuses to let him use that word and threatens to hang up on him. This incident enacted a terminology battle on the micro-level, which is where such battles often have dramatic effects on people's everyday lives.

113. Norman L. Geisler and J. Kerby Anderson, *Origin Science: A Proposal for the Creation-Evolution Controversy* (Grand Rapids, Mich.: Baker Book House, 1987).

114. Keith, *Scopes II*, vii–viii. Subsequent page references appear in parentheses in the text. This quotation from Darrow was also cited in Geisler's *Creator in the Courtroom* but was not used as an organizing motif.

115. See summaries in Geisler, *Creator in the Courtroom*, 121, 126.

116. These facts are summarized most succinctly in ibid., 24–25.

117. Keith, *Scopes II*, 116–117.

118. Ecker, *Dictionary of Science and Creationism*, 135.

119. Facts about Judge Overton are drawn from his obituary in the *New York Times* of July 14, 1987, B6. Overton died of cancer at the tragically young age of 47.

120. Geisler, *Creator in the Courtroom*, 47.

121. *McLean v. Arkansas Board of Education* 529 F.Supp. 1255 (E.D. Ark., 1982). Further page references to this opinion appear in parentheses in the text.

122. David S. Caudill, "Law and Worldview: Problems in the Creation-Science Controversy," *Journal of Law and Religion* 3 (1985): 1–46. Judge Overton's decision has occasioned many other legal commentaries besides Caudill's including the following essays: William H. Becker, "Creationism: New Dimensions of the Religion-Democracy Relation," *Journal of Church and State* 27 (1985): 315–333; Lucien J. Douge, "From *Scopes* to *Edwards*: The Sixty-Year Evolution of Biblical Creation in the Public School Curriculum," *University of Richmond Law Review* 22 (1987): 187–234; Robert M. Gordon, "*McLean v. Arkansas Board of Education*: Finding the Science in 'Creation Science,'" *Northwestern Law Review* 77 (1982): 374–402; Gary C. Leedes, "Monkeying Around with the Establishment Clause and Bashing Creation-Science," *University of Richmond Law Review* 22 (1987): 149–186; Wayne V. McIntosh, "Litigating Scientific Creationism, or 'Scopes' II, III, . . . ," *Law and Policy* 7 (1985): 378–394; Morrell E. Mullins, "Creation Science and *McLean v. Arkansas Board of Education*: The Hazards of Judicial Inquiry into Legislative Purpose and Motive," *University of Arkansas at Little Rock Law Journal* 5 (1982): 345–396; Nadine Strossen, "'Secular Humanism' and 'Scientific Creationism': Proposed Standards for Reviewing Curricular Decisions Af-

fecting Students' Religious Freedom," *Ohio State Bar Journal* 47 (1986): 333–407; and Judith A. Villarreal, "God and Darwin in the Classroom: The Creation/Evolution Controversy," *Chicago-Kent Law Review* 64 (1988): 335–374.

123. Larson *Trial and Error*, 156–157. Larson indicates that 70 percent of the respondents favored teaching both accounts, 10 percent favored teaching evolutionism only, 10 percent favored teaching creationism only, and 10 percent expressed no opinion. In two interesting exceptions however, over one-third of evangelicals and fundamentalists responded that only creationism should be taught, and over two-thirds of college professors responded that only evolutionism should be taught. Apparently, college professors oppose creationism more firmly than fundamentalists oppose evolutionism.

124. Jim Merkel, "Judge Who Ruled against Creation Law Speaks Out," *Christianity Today* 26.19 (1982): 48, 50, 52.

125. William R. Overton, "Speech to Pennsylvania Appellate Judges, Bucknell University, Lewisburg, Pa., July 29, 1982 (typed ms.), Law Clerk's File, *McLean v. Arkansas Board of Education* 529 F. Supp. 1255 (E. D. Ark. 1982). 6, 17–18. Subsequent page references to this speech appear in parentheses in the text.

5

Conclusion

Garry Wills begins his 1990 book, *Under God: Religion and American Politics,* as follows: "The learned have their superstitions, prominent among them a belief that superstition is evaporating. Since science has explained the world in secular terms, there is no more need for religion, which will wither away. Granted, it has been slow to die in America."[1] If there is a recurring lesson in each episode of the creation/evolution controversy, it is that neither creationism nor evolutionism ever dies in America. This controversy does not go away—no matter how convinced one side is that truth has won out and the other side, that error has prevailed. Wills could have been speaking of the creationists when he reminds his readers that "in a time of reviving fundamentalisms [around the world], some Americans have rediscovered our native fundamentalists (a recurring, rather than cumulative experience for the learned). It seems careless for scholars to keep misplacing such a large body of people."[2]

This large body of people is not convinced of the truth of evolution, nor do they plan to remain silent about their opposition to it. In the wake of the 1987 U.S. Supreme Court ruling that creationism cannot be legislated into public schools, they have not conceded the debate. Instead of continuing to pass state laws, they have developed new battle tactics, focusing in the 1990s especially on debates in local school districts. They will probably keep up the fight until they can join Andrew Young and the former Communists of whom he was speaking in 1989 in a fundamentalist victory song: "When they come out from behind the Iron Curtain, they are singing 'We Shall Overcome,' a Georgia Baptist hymn."[3]

Perhaps this controversy has continued to fascinate authors for over a century because of the unrelenting and passionate commitments it inspires on both sides. Ronald L. Numbers begins his 1992 award-winning history, *The Creationists*, as follows: [The] history of modern creationism includes some of the fiercest skirmishes in the annals of science and religion." He concludes: "Its shocking success, limited though it may have been, shattered facile beliefs about the inevitability of secularization and scientific progress and called into question long-cherished convictions about the relationship between science and religion."4 Foremost among these convictions are the notion that scientific knowledge will gradually displace religious belief and that teaching the truth about evolution will diminish belief in the errors of creationism. Despite the highest hopes of evolutionists, creationists have failed to see the light, and vice versa. In each confrontation we have traced, both sides seem unable to see the other side's light, apparently preferring to remain in darkness. It makes one wonder if light is the best metaphor to use.

This book has argued that the creation/evolution controversy does not finally make sense if one imagines it as a clash between light and darkness, truth and error. Both sides have inherited such a vocabulary and used it throughout the episodes we have traced, but it has only embroiled them in shouting matches about who is defining key terms correctly, with each side convincing the appropriate governing body (legislature, court, scientific community) to police the appropriate definitions for a time—but only for a time. Within our current legal system and our culture of disciplinary communities, defining key terms is, finally, a political act influenced by the basic values and beliefs of the definer. It is time to admit that defining is political and to analyze its politics, asking if we are satisfied as a culture to arrive at our definitions of important terms by asking experts on the basis of their specialized knowledge. Do we want an expert from one discipline to proclaim the "correct" definition, only to be challenged by another expert from another discipline with another "correct" definition? Do we want to keep arguing about who is telling the truth and who is lying, to perpetuate our endless contemporary battle between rhetoric and philosophy?

I have an alternate suggestion. It is based on my major contention that values and beliefs are more basic than truths. In the world as we know it, we have no simple way of telling what is truth. "What is truth?" asked Pontius Pilate of Jesus Christ, and the New Testament indicates that even Christ did not directly answer (John 18:38). If, instead of seeing truth as distinct from either values or beliefs, we begin to see it as a function of shared values and beliefs, as the name we give to that which seems obvious to us within our various interpretive communities, I believe we can begin to untangle some confusions and to reason more effectively with others who do not share our worldviews. A major argument of this book has been that liberal political philosophy itself is a set of commitments not fully shared by fundamentalists. In effect, within a conception of language as shared human perceptions and commitments that we are attempting to articulate in words, every argument becomes an argument about values and beliefs, including those arguments made in the name of truth as the highest value by those who believe (with much of our culture) that truth can be clearly and easily discerned.

Using this rhetorical conception of language, how can one best account for the creation/evolution controversy? I believe it is better to see this controversy as a contest between values and beliefs than as a battle between truth and error. Throughout this book I have argued that the two sides represent different worldviews competing for the power to represent reality itself in the culture as a whole. Let me spell out the political implications of this thesis by considering ideas from several recent works about creationism.

A first idea is that this controversy finally reflects a deep conflict between the basic tenets of liberalism and fundamentalism. In his essay, "Evolutionism, Creationism, and Treating Religion as a Hobby," Stephen D. Carter writes: "the liberal believes that reason is the most important human faculty, and that amenability to reason is the trait that distinguishes humans from the rest of creation."[5] The problem arises when a fundamentalist challenges this "faith in the faculty of reason" (988) by positing instead a faith in "God's revelation; no artifice of mortal man can contradict that; and any 'evidence' that the revelation is incorrect is either erroneous or deceptive" (993). Carter believes that by valuing faith in the Bible more than faith in reason itself as a primary criterion for determining truth, fundamentalists effectively reject the basis of liberalism: "They are independent thinkers who insist on a right to their own means for seeking knowledge of the world, and they deny the right of the state to tell their children that their worldview is wrong" (981). For Carter, this clash of worldviews leads liberalism to tolerate religious beliefs only by removing them from the public sphere to the private. In his book *The Culture of Disbelief*, Carter further focuses on the problem of dealing with religious beliefs in a culture erected on Enlightenment ideals. Identifying again a key conflict between worldviews, he writes, "Not everyone agrees that the Enlightenment project of replacing divine moral authority with the moral authority of human reason was a good idea." In the creation/evolution controversy, he concludes, one can see a "war . . . between competing systems of discerning truth."[6]

This basic challenge to liberal political philosophy is further illuminated in a provocative response to Carter's essay by Stanley Fish. Fish writes:

> Liberalism is informed by a faith (a deliberately chosen word) in reason as a faculty that operates independently of any particular world view. It is therefore committed at once to allowing competing world views equal access to its deliberative arena, and to disallowing the claims of any one of them to be supreme, unless of course it is demonstrated to be at all points compatible with the principles of reason. . . . However, if you take away the "primacy of reason," liberal thought loses its integrity, has nothing at its center, becomes just one more competing ideology rather than a procedure (and it is in procedure or process that liberalism puts its faith) that outflanks or transcends ideology. . . . Indeed, liberalism depends on not inquiring into the status of reason.[7]

Fish argues that liberalism depends on such Enlightenment notions as universal reason and justice to decide between competing positions. However, these notions have increasingly come under attack. Indeed, many other poststructuralists (among them Jacques Derrida, Michel Foucault, Alasdair MacIntyre, and writers not discussed in this book, such as Mikhail Bakhtin, Barbara Herrnstein Smith, Stanley Hauerwas, and Roberto Unger) have also concluded that reason and justice simply cannot be *universal* but are instead *contingent*, *dialogic*, *economic*, or, in the account developed here, *rhetorical*: they are words used differently by people in various contexts to do various kinds of persuasive work.[8] By judging between competing worldviews on the basis of rationality, liberalism effectively disqualifies any worldview that does not agree on rationality's primary value.

Evolutionists clearly share this estimation of rationality and enter public discussions in order to defend it. They see creationism as an attack on this value and thus call for its rejection from the public schools. So do members of other academic disciplines (as described in Chapter 4); they come to the aid of evolutionists in defense of the same value. However, creationists do not share this key value. They argue for a different one: having faith in the Bible instead. According to the laws of liberal dialogue, they thus ought to be eliminated from the discussion before it begins. However, the creationists refuse to remain silent. They argue for their own values rather than the central value of liberalism as a political philosophy. They do not see their beliefs as private and subjective, and thus do not agree with liberals that beliefs ought to be kept out of the political arena. Their sense of the rightness of their beliefs leads them to try to defend them in the public forum against all odds, if necessary by inventing terms like *creation-science* and enacting their inclusion through law. The legislative efforts of the creationists directly attack another central belief of liberalism, the notion that religious beliefs ought not to influence public laws. To use Fish's terms, liberalism itself is a belief that a political system can be strictly rational, free of beliefs that would interfere with the neutral workings of reason and justice in a body politic. However, if one rejects these claims for objective and impersonal reason, Fish holds that liberalism has nothing to recommend it as a political philosophy. Thus, concludes that liberalism itself does not exist in a coherent form.

Although I have argued throughout this book that liberal political philosophy involves such logical inconsistencies as those pointed out by Carter and Fish, I have not meant to imply that liberalism should be rejected. All positions articulated in language involve such blind spots (which they take for granted), because language itself is an imperfect instrument for thought, a rhetorical tool used by human communities and not a positivist namer of impersonal and timeless realities. Instead of either supporting or opposing liberalism or fundamentalism, I have attempted to show that evolutionists and creationists disagree on the shape of the world, not just its biological and geological history. They argue publicly in an effort to convince the culture as a whole to see the world that they see.

This major disagreement at the level of perception is further articulated by Raymond A. Eve and Francis B. Harrold in their comprehensive book, *The Creationist Movement in Modern America*:

> Far from simply debating the scientific evidence, it appears that creationist and evolutionist groups structure their perceptions of reality in very different ways, based on very different cognitive principles and on different assumptions about the rules of knowing. Thus any social analysis of the movements involved will deal less with natural science and much more with the social psychology and worldviews of differing human groups.[9]

Eve and Harrold's book is an extended treatment of these differences in worldviews. Later in the work, they explain: "Creationists and their opponents tend to differ not over competing theories within the same intellectual framework, but in their most profound understandings of reality, religion, American society, and the nature of the scientific enterprise. Given these differences, it is extremely unlikely that the creation-evolution controversy will end with creationists being persuaded that mainstream scientists are correct (or vice versa)" (67). In the view of Eve and Harrold, the debate will continue indefinitely as each side attempts to convince the culture as whole to share its perspectives. As a result of this endless battle, they ask another major question: "Who arbitrates what is valid knowledge in society? What, it might be asked, is the proper role of scientific authority and its associated establishment when it conflicts with popular opinion in a democratic society?" (10) This volume has argued that American culture has not resolved this problem, but instead now pits disciplinary experts against other disciplinary experts, all speaking in the name of truth. A final quotation from Eve and Harrold indicates what these experts are doing: "Each side perceives and presents itself as defending essential civilized virtues from the other's assault" (61). This controversy is ultimately a debate about who gets to clothe virtue in the language of truth.

In naming the two worldviews that are at war in the creation/evolution controversy, Eve and Harrold borrow their key terms from the work of two other sociologists, Ann L. Page and Donald A. Clelland. In a study of a 1974 textbook controversy in Kanawha County, West Virginia, Page and Clelland trace a battle between two "status groups," which they call the "cultural modernists" and the "cultural fundamentalists": "A status group stands for a way of life; and such groups are consequently involved in constant struggles for control of the means of symbolic production through which their reality is constructed."[10] In fighting bitterly about what textbooks should be used to teach their children, participants on both sides debated the same basic issues as the creationists and the evolutionists. On one side were people who supported rational inquiry on the basis of secular evidence; on the other side were people who opposed the widespread attacks on God, the Bible, and other key values in their own culture that they saw in the disputed textbooks. This conflict erupted into violence and resulted in major political upheavals, including a near-secession of part of the county and the founding of many Christian schools. Parallel to the creation/evolution controversy, the Kanawha textbook controversy finally turned on a battle between worldviews, between conflicting "attempts to build and sustain moral orders which provide basic meaning for human lives."[11]

The creation/evolution controversy is, finally, a similar conflict between moral orders. In a book entitled *Darwin on Trial*, the most powerful recent attack on evolution, University of California at Berkeley law professor Phillip E. Johnson attempts to discredit evolution as "the scientific orthodoxy of today" by investigating the question of whether the theory of evolution "is based upon a fair assessment of the scientific evidence, or whether it is another kind of fundamentalism."[12] Although I disagree with his ultimate conclusion and think it is impossible for a nonscientist to assess such evidence fairly, Johnson also sees the conflict at the level of worldview:

Darwinist evolution is an imaginative story about who we are and where we came from, which is to say it is a creation myth. As such, it is an obvious starting point for speculation about how we ought to live and what we ought to value. A creationist appropriately starts with God's creation and God's will for man. A scientific naturalist just as appropriately starts with evolution and with man as a product of nature.

In its mythological dimension, [the theory of evolution] is the story of humanity's liberation from the delusion that its destiny is controlled by a power higher than itself. (131)

Although some evolutionists might reject this redaction of their creation myth, Johnson helpfully identifies the power of both stories as stories. I take the creation/evolution controversy partly as a battle between two stories that both have captured human hearts, suggesting what is good about humans, how they should organize their social lives, and what their duty ought to be. Both stories have persuaded thousands of hearers, many of whom have worked tirelessly to help their own story achieve a wider hearing and have a greater impact on others' lives. Each story reflects some basic values and beliefs of its tellers, which it simultaneously reinforces. I agree with Johnson on the larger struggle encoded in this controversy in American culture at the present time: "The leaders of science see themselves locked in a desperate battle against religious fundamentalists, a label which they tend to apply broadly to anyone who believes in a Creator who plays an active role in worldly affairs. These fundamentalists are seen as a threat to liberal freedom. . . . As the creation myth of scientific naturalism, [the theory of evolution] plays an indispensable ideological role in the war against fundamentalism" (153). Anyone who has witnessed American politics since 1980 cannot doubt that fundamentalism and its opponents are indeed involved in a war. However, I think Johnson perceives the conflict too narrowly. On the side of the evolutionists are not just atheistic scientists, but many other people from a wide variety of persuasions who want to defend the political system known as liberalism from a variety of assaults around the nation and the world. In this controversy, both sides are defending deep commitments; both fear the consequences of losing this war.

Soon after Johnson's book appeared in print, it was reviewed by Stephen Jay Gould for *Scientific American* in an article entitled, "Impeaching a Self-Appointed Judge." Initially welcoming any insights from a lawyer as "enlightenment that intelligent outsiders could bring to the puzzles of a discipline," Gould contends that "to be useful in this way, a lawyer would have to understand and use our norms and rules, or at least tell us where we err in our procedures; he cannot simply trot out some applicable criteria from his own world and falsely condemn us from a mixture of ignorance and inappropriateness. Johnson, unfortunately, has taken the low road in writing a very bad book entitled *Darwin on Trial*."[13] Gould methodically proves that Johnson does not have the scientific expertise to quarrel with evolution, but he also finally repeats the pattern I have been attacking throughout this study: he argues that Johnson makes "occasional use of scientific literature only to score rhetorical points" (119) and "tries to impeach us by rhetorical trickery," adding, "no wonder lawyer jokes are so popular in our culture" (120). Gould sees Johnson's rejection of evolution as ignorance and blindness and accuses him of following all the lawyers and rhetoricians throughout history in lying to attack the truth. I have argued that the truth is not as obvious as Gould believes. In this review in a leading popular science journal, Gould finally uses his own rhetoric to counter the rhetoric of Phillip Johnson. He ends by calling Johnson's book "an acrid little puff" that does not even begin to compare to the power of *Inherit the Wind* (121). In this final allusion, Gould refers to his own favorite story of the Scopes "Monkey" Trial to support his position, just as Johnson predicted in calling this controversy a battle between opposing stories.

Creationism is finally only one of many battles in American politics that is based on such larger conflicts. In his book *Culture Wars*, James Davison Hunter writes about such conflicts as follows: "[T]he culture war emerges over fundamentally different conceptions of moral authority, over different ideas and beliefs about truth, the good, obligation to one another, the nature of community, and so on. It is, therefore, cultural conflict at its deepest level."[14] Although people often see such conflict at the level of race, class, ethnic group, or the meaning of America itself, Hunter concludes: "But in the final analysis, whatever else may be involved, cultural conflict is about power—a struggle to achieve or maintain the power to define reality."[15]

I see in the creation/evolution controversy a battle for cultural power and an example of one great battle of our generation. This recurring debate about the evidence for creation or evolution, about whose viewpoint is the plain truth and whose is deceptive rhetoric, raises some very important questions. Will our nation dedicate itself to humanity or to God? Should our public schools teach skeptical doubt or trusting faith in some other ideal? Should we defer to science or religion when their answers to particular questions conflict—or should we try to show how both answers make sense in different arenas? When we celebrate the great achievements of the past, should we tell stories of heroic scientists or miraculous events from the Bible? These are some of the stakes, as is the case in many other contemporary battles. I finally agree more with the evolutionists than the creationists, but I do not want the creationists to give up the fight. I am increasingly convinced that reason and knowledge are not the only bases on which to found a society, nor even that they are the best. I am unsure that a strictly rational society is best. Truth is good, but what about justice, compassion, mercy, and love? How does one found a society on these values? I doubt that either the creationists or the evolutionists will ever stop arguing so long as we have no simple way to know the truth beyond own our perceptions. We in this pluralistic nation have had to continually deal with recurring tensions between professionalism and democracy, between the academy and other cultural institutions, between competing political philosophies, and between other differing persuasions, all arguing for, and from within, their own worldviews. In this life, we walk by faith. We must put our faith in those persuasions that seem most worthy of it.

How can we make sense of it all? Perhaps the best way is to stop trying to understand it rationally and return to the telling of stories. Over the years of pondering this controversy, I have also been teaching students about the power of stories and have found three especially that speak to my heart. The first is a delightful novel by Fred Chappell about a high school science teacher in western North Carolina; it brilliantly enacts the creation/evolution controversy as a function of human fears and misunderstandings.[16]

The novel follows the main character, Joe Robert Kirkman, through a very eventful day in his life. On this "balmy May Friday in 1946" (9), Joe Robert must appear before the local school board to answer complaints by a fundamentalist student's parents that he is teaching their son to disbelieve the Bible and instead accept the theory of evolution. At the appointed time, Joe Robert, feeling overcome with fear, sticks his head in the door of the classroom where the school board is waiting and yells, "You can't fire me; I quit!" (170). Then he goes to hide in the bathroom. In his absence, the six board members have a hilarious conversation in which they all try to figure out what Joe Robert said. They reveal that they had no intention of firing him; they were only meeting with him to appease the parents. One member says, "As far as I'm concerned, Joe Robert can come in here right now and shake hands with us and then we can all go home" (175). Back in the bathroom, Joe realizes his mistake: "I don't need to behave like a madman. . . . Those are my friends sitting in those school desks. They're reasonable people; we could have worked something out. I know they don't want me to lose my job" (179). However, he fails to summon up enough courage to return to the classroom until the board members have all left. When Joe Robert enters the empty room, Chappell brilliantly symbolizes the fears and frustrations at the heart of this controversy in the sound of the clock: "The black minute hand clicked sharply as it moved to 3:36. That noise sounded . . . like a guillotine blade coming down thwack" (182).

Through a clever twist in the plot based on previous incidents, everything ultimately turns out fine. Because he had saved the governor's grandniece from drowning early in the morning, that very afternoon Joe Robert gets a job as the head of the Governor's Special Commission on Education, where he can work to advance science education throughout the state. When he returns home to his family, at the end of the novel, the reader realizes that Chappell has been telling his own heroic creation myth; the narrator-son reveals the major plot pattern of the novel in "a dream of my father as Aeneas as he descended into the underworld to meet the dead and rose into the light to talk with the gods and battled the backward barbarian forces so that civilization might find a foothold in a scrannel and unpromising soil" (201–202). By telling the story of a North Carolina high school science teacher who is like Aeneas in founding a new civilization, Chappell clearly reveals some of his own key values (education, courage, endurance, humor), but only after showing throughout the novel that he understands a creationist's viewpoint of this controversy better than any other author I have read. I applaud his compassion for the creationists at the same time that he boldly rejects them in the character of Joe Robert and represents the beliefs that matter most to him. Chappell clearly teaches his own fundamental commitments without rejecting or ridiculing those of the creationists.

The second story is a brief allegory about truth given by John Milton in the *Areopagitica*, his great imaginary speech to the British House of Lords and Commons in defense of a free press:

> Truth indeed came once into the world with her divine Master, and was a perfect shape most glorious to look on: but when he ascended, and his Apostles after him were laid asleep, then strait arose a wicked race of deceivers, who as that story goes of the AEgptian Typhon with his conspirators, how they dealt with the good Osiris, took the virgin Truth, hewd her lovely form into a thousand peeces, and scatter'd them to the four winds. From that time ever since, the sad friends of Truth, such as durst appear, imitating the careful search that Isis made for the mangl'd body of Osiris, went up and down gathering up limb by limb still as they could find them. We have not yet found them all, Lords and Commons, nor ever shall doe, till her Masters second comming; he shall bring together every joynt and member, and shall mould them into an immortall feature of loveliness and perfection.17

In an inimitable style that reshapes classical allusions for present purposes, Milton compares Truth to a body "of perfect shape" that was "hewed
into a thousand pieces" and now must be gathered and put together by her
"sad friends." They are sad in part because they can only see her full body
now with an eye of faith; they must wait to find the rest of her lost limbs and
to see her "brought together" into a form of "loveliness and perfection" until
"her Masters second comming." In this allegory, Milton's conception of
truth complements, instead of competing with, his knowledge of rhetoric.
He suggests that in this imperfect world we all must wait, "gathering up" the
pieces of truth and hoping for that brighter day when she shall appear in all
her glory and we will no longer need to rely on persuasion. Perhaps both the
creationists and the evolutionists could learn about truth and error from
John Milton. Indeed, we all might learn to wait and hope with him.

My final story is really just a passage from *All the King's Men* (1946), a
compelling novel by Robert Penn Warren about a ruthless but persuasive
political leader in the American South during the 1930s. In this passage,
Willie Stark, a self-taught country lawyer who has become a powerful governor, explains his conception of the law:

> "No," the Boss corrected, "I'm not a lawyer. I know some law. In fact, I know a
> lot of law. And I made me some money out of law. But I'm not a lawyer. That's
> why I can see what the law is like. It's like a single-bed blanket on a double
> bed and three folks in the bed and a cold night. There ain't ever enough blan-
> ket to cover the case, no matter how much pulling and hauling, and some-
> body is always going to nigh catch pneumonia. Hell, the law is like the pants
> you bought last year for a growing boy, but it is always this year and the seams
> are popped and the shankbones to the breeze. The law is always too short and
> too tight for growing humankind. The best you can do is do something and
> then make up some law to fit and by the time that law gets on the books you
> would have done something different."[18]

Throughout my study of the creation/evolution controversy and my own
lengthening thread of analysis, I have felt as if something in American culture were being stretched too tight, as if the blanket of American law and its
underlying sheet of liberal political philosophy did not quite fit for issues
such as this. For 140 years creationists and evolutionists alike have been using various strategies to attempt to jerk the bedcovers over to their side of
the bed. Thus, they have taken turns being warmly covered and then shiv-
ering through a cold and windy night. However, for those who finally see
this controversy as a battle for cultural power pitting humanists against fundamentalists, the controversy involves more than a struggle to see who ends

up with the covers on a particular night. Perhaps it suggests that the covers themselves need to stretch farther. Perhaps we need to think longer and harder about the relationship between knowledge and politics, broaden our vision of how to be just and fair, and figure out how to teach what we value and believe in a way that longs for truth but does not simply impose our own visions of it. Perhaps our culture needs to weave some new cloth into the bedcovers of our law and our political philosophy—to make sense of both the rhetorical warp and the philosophical woof.

NOTES

1. Garry Wills, *Under God: Religion in American Politics* (New York: Simon and Schuster, 1990), 15.

2. Ibid.

3. Ibid.

4. Ronald L. Numbers, *The Creationists* (New York: Alfred A. Knopf, 1992), xiv, 339.

5. Stephen D. Carter, "Evolutionism, Creationism, and Treating Religion as a Hobby," *Duke Law Journal* 1987: 987. Further page references appear in parentheses in the text.

6. Stephen D. Carter, *The Culture of Disbelief* (New York: Basic Books, 1993), 225, 176.

7. Stanley Fish, "Liberalism Doesn't Exist," *Duke Law Journal* 1987: 997.

8. [Mikhail Bakhtin], V. N. Voloshinov, *Marxism and the Philosophy of Language*, trans. Ladislav Matejka and I. R. Titunik (Cambridge: Harvard University Press, 1986); Barbara Herrnstein Smith, *Contingencies of Value* (Cambridge: Harvard University Press, 1988); Stanley Hauerwas, *Christian Existence Today* (Durham, N.C.: Labyrinth Press, 1988); Robert Mangabeira Unger, *Knowledge and Politics* (New York: Macmillan/Free Press, 1975, 1984).

9. Raymond A. Eve and Francis B. Harrold, *The Creationist Movement in Modern America* (Boston: Twayne, 1991), 6. Subsequent page references appear in parentheses in the text.

10. Ann L. Page and Donald A. Clelland, "The Kanawha County Textbook Controversy: A Study of the Politics of Lifestyle Concern," *Social Forces* 57 (1978): 266.

11. Ibid., 279.

12. Phillip E. Johnson, *Darwin on Trial* (Washington, D.C.: Regnery Gateway, 1991), 3, 14. Subsequent page references appear in parentheses in the text.

13. Stephen Jay Gould, "Impeaching a Self-Appointed Judge," book review of *Darwin on Trial* by Phillip E. Johnson, *Scientific American* 267 (July 1992): 118–119. Subsequent page numbers appear in parentheses in the text.

14. James Davison Hunter, *Culture Wars* (New York: Basic Books, 1991), 49.

15. Ibid., 52.

16. Fred Chappell, *Brighten the Corner Where You Are* (New York: St. Martin's Press, 1989). Subsequent page references appear in parentheses in the text.

17. John Milton, "Areopagitica." *The Prose of John Milton*, ed. J. Max Patrick (New York: New York University Press, 1968), 316–317. The original spelling, italics, and punctuation were retained.

18. Robert Penn Warren, *All the King's Men* (New York: Harvest/HBJ, 1946), 136. This novel is loosely based on the notorious career of Louisiana Governor Huey Long.

References

Adler, Jerry. "Enigmas of Evolution." *Newsweek*, 29 March 1982: 44–49.

Allen, Leslie H. *Bryan and Darrow at Dayton*. New York: Arthur Lee, 1925.

American Association for the Advancement of Science. *Evolutionists Confront Creationists*. San Francisco: Author, 1984.

Ashby, LeRoy. *William Jennings Bryan: Champion of Democracy*. Boston: Twayne, 1987.

Asimov, Isaac. "The 'Threat' of Scientific Creationism." In *Science and Creationism*, ed. Ashley Montagu, 182–193. New York: Oxford University Press, 1984.

Bailey, Kenneth K. "The Enactment of Tennessee's Anti-Evolution Law." *Journal of Southern History* 16 (1950): 472–490.

[Bakhtin, Mikhai]. V. N. Voloshinov. *Marxism and the Philosophy of Language*. Trans. Ladislav Matejka and I. R. Titunik. Cambridge: Harvard University Press, 1986.

Barr, James. *Fundamentalism*. Philadelphia: Westminster Press, 1978.

Barrett, Paul H., Donald J. Weinshank, and Timothy T. Gottleber, eds. *A Concordance to Darwin's "Origin of Species," First Edition*. Ithaca, N.Y.: Cornell University Press, 1981.

Barrett, Paul H., Donald J. Weinshank, Paul Ruhlen, and Stephan J. Ozminski, eds. *A Concordance to Darwin's "The Descent of Man, and Selection in Relation to the Sexes."* Ithaca, N.Y.: Cornell University Press, 1987.

Becker, William H. "Creationism: New Dimensions of the Religion-Democracy Relation." *Journal of Church and State* 27 (1985): 315–333.

Birchler, Allen. "The Anti-Evolutionary Beliefs of William Jennings Bryan." *Nebraska History* 54 (1973): 545–559.

Bird, Wendell R. "Evolution in Public Schools and Creation in Students' Homes: What Creationists Can Do: Parts I and II." In *The Decade of Creation*, ed. Henry M. Morris and Donald H. Rohrer, 119–131. San Diego: Creation-Life Publishers, 1981.

———. "Freedom from Establishment and Unneutrality in Public School Instruction and Religious School Regulation." *Harvard Journal of Law and Public Policy* 2 (1979): 125-205.

———. "Freedom of Religion and Science Instruction in Public Schools." *Yale Law Journal* 87 (1978): 515–570.

———. *The Origin of Species Revisited*. 2 vols. New York: Philosophical Library, 1994.

Boles, Donald E. "Religion in the Schools: A Historical and Legal Perspective." In *Did the Devil Make Darwin Do It?* ed. David B. Wilson, 170–188. Ames: Iowa State University Press, 1983.

Bolton, S. Charles. "The Historian as Expert Witness: Creationism in Arkansas." *Public Historian* 4 (1982): 59–67.

Boorstin, Daniel J. *The Image: A Guide to Pseudo-Events in America*. New York: Atheneum, 1961, 1985.

Bowler, Peter J. "The Changing Meaning of 'Evolution.'" *Journal of the History of Ideas* 36 (1975): 95–114.

———. *The Non-Darwinian Revolution: Reinterpreting a Historical Myth*. Baltimore: Johns Hopkins University Press, 1988.

Bryan, Mary Baird. *The Memoirs of William Jennings Bryan*. Philadelphia: John C. Winston, 1925.

Bryan, William Jennings. *The Bible and Its Enemies*. Chicago: Bible Institute Colportage Association, 1921.

———. "Man." In *Speeches of William Jennings Bryan* 2:291–314. New York: Funk and Wagnalls, 1913.

———. "The Menace of Darwinism." In *In His Image*, 86–135. New York: Fleming H. Revell, 1922.

———. "The Origin of Man." In *Seven Questions in Dispute* 133–158. New York: Fleming H. Revell, 1924.

———. "The Prince of Peace." In *Speeches of William Jennings Bryan* 2:261–290. New York: Funk and Wagnalls, 1913.

———. "The Spoken Word." In *In His Image*, 248–266. New York: Fleming H. Revell, 1922.

———, ed. *The World's Most Famous Orations*. 10 vols. New York: Funk and Wagnalls, 1906.

Buckley, Michael J. *At the Origins of Modern Atheism*. New Haven: Yale University Press, 1987.

Campbell, John Angus. "Charles Darwin: Rhetorician of Science." In *The Rhetoric of the Human Sciences*, eds. John S. Nelson, Alan Megill, and Donald N. McCloskey, 69–86. Madison: University of Wisconsin Press, 1987.

————. "The Invisible Rhetorician: Charles Darwin's Third Party Strategy." *Rhetorica* 7 (1989): 55-85.

Carter, Paul. *The Twenties in America*. New York: Thomas Y. Crowell, 1968.

Carter, Stephen L. *The Culture of Disbelief*. New York: Basic Books, 1993.

————. "Evolutionism, Creationism, and Treating Religion as a Hobby." *Duke Law Journal* 1987: 977–996.

Caudill, David S. "Law and Worldview: Problems in the Creation-Science Controversy." *Journal of Law and Religion* 3 (1985): 1–46.

Chappell, Fred. *Brighten the Corner Where You Are*. New York: St. Martin's Press, 1989.

Chittick, Donald E. *The Controversy: Roots of the Creation/Evolution Conflict*. Portland, Ore.: Multnomah Press, 1984.

Chomsky, Noam. *Language and Problems of Knowledge: The Managua Lectures*. Cambridge: MIT Press, 1988.

Coffin, Harold G. *Origin by Design*. Washington, D.C.: Review and Herald Publishing Association, 1983.

Conley, Thomas J. *Rhetoric in the Western Tradition*. New York: Longman, 1990.

Corbett, Edward P. J. "A Survey of Rhetoric." In *Classical Rhetoric for the Modern Student*, 535–568. New York: Oxford University Press, 1965.

"Creationism and Evolution." *Zygon: Journal of Religion and Science* 22.2 (special issue) (June 1987).

"The Creationist Attack on Science." *Federation of American Societies for Experimental Biology Proceedings* 42 (1983): 3022–3042.

Creation-Science Fellowship. *Proceedings of the First International Conference on Creationism*. 2 vols. Pittsburgh: Author, 1986.

Darrow, Clarence. *The Story of My Life*. New York: Charles Scribner's Sons, 1932.

Darwin, Charles. *Calendar of the Letters of Charles Robert Darwin to Asa Gray*. Boston: Historical Records Survey, 1939.

————. *More Letters of Charles Darwin*. ed. Francis Darwin. 2 vols. London: John Murray, 1903.

————. *On the Origin of Species, 1859*. Vol. 15, of *The Works of Charles Darwin*, ed. Paul H. Barrett and R. B. Freeman. London: William Pickering, 1988.

Dean, Michael P. "Language and Character Formation in *Inherit the Wind*." *Publications of the Arkansas Philological Association* 8 (1982): 16–23.

Dixon, Peter. *Rhetoric*. London: Methuen, 1971.

Douge, Lucien J. "From *Scopes* to *Edwards*: The Sixty-Year Evolution of Biblical Creation in the Public School Curriculum." *University of Richmond Law Review* 22 (1987): 187–234.

Duffy, Susan. "The Origin of Speeches: *Inherit the Wind*, Irving Stone, and the Scopes Trial." *American Notes and Queries* 22 (1983): 14–17.

Durant, John. "Darwinism and Divinity: A Century of Debate." In *Darwinism and Divinity: Essays on Evolution and Religious Belief*, ed. John Durant, 9–39. New York: Basil Blackwell, 1985.

Ecker, Ronald L. *Dictionary of Science and Creationism*. Buffalo, N.Y.: Prometheus Press, 1990.

[Edwords, Frederick.] "Letter to the Reader." *Creation/Evolution*, Issue 1 (1980): i.

Eger, Martin. "A Tale of Two Controversies: Dissonance in the Theory and Practice of Rationality." *Zygon* 23 (1988): 291–368.

Ehrenfeld, David. *The Arrogance of Humanism*. New York: Oxford University Press, 1978.

Eldredge, Niles. *The Monkey Business: A Scientist Looks at Creationism*. New York: Washington Square Press, 1982.

Ellis, William E. "Evolutionism, Fundamentalism, and the Historians: A Historiographical Review." *Historian* 44 (1981): 15–35.

Eve, Raymond A., and Francis B. Harrold. *The Creationist Movement in Modern America*. Boston: Twayne, 1991.

"The Evolution-Creation Science Controversy." *College Board Review*, no. 123 (special issue) (spring 1982). "Evolution vs. Creationism: The Schools as Battleground." *Humanist* 37.1 (special issue) (January/February 1977).

Feyerabend, Paul. *Against Method*. Rev. ed. London: Verso, 1988.

Fish, Stanley. *Doing What Comes Naturally: Change, Rhetoric, and the Practice of Theory in Literary and Legal Studies*. Durham, N.C.: Duke University Press, 1989.

———. *Is There a Text in This Class? The Authority of Interpretive Communities*. Cambridge: Harvard University Press, 1980.

———. "Liberalism Doesn't Exist." *Duke Law Journal* 1987: 997–1001.

———. *Surprised by Sin: The Reader in "Paradise Lost."* Berkeley: University of California Press, 1967.

Foucault, Michel. *The History of Sexuality. Vol. 1, An Introduction*. Trans. Robert Hurley. New York: Random House/Vintage, 1980.

Frair, Wayne, and Percival Davis. *The Case for Creation*. Chicago: Moody Press, 1983.

Frye, Roland Mushat, ed. *Is God a Creationist? The Religious Case against Creation-Science*. New York: Charles Scribner's Sons, 1983.

Futuyma, Douglas J. *Science on Trial: The Case for Evolution*. New York: Pantheon, 1983.

Gates, Henry Louis, Jr. *The Signifying Monkey: A Theory of Afro-American Literary Criticism*. New York: Oxford University Press, 1988.

Gatewood, Willard B., Jr. "From Scopes to Creation-Science: The Decline and Revival of the Evolution Controversy." *South Atlantic Quarterly* 83 (1984): 363–383.

———. Review of Larson's *Trial and Error*. *Georgia Historical Quarterly* 70 (1986): 371–373.

Geisler, Norman L., *The Creator in the Courtroom*. Milford, Mich.: Mott Media, 1982.

Geisler, Norman L., and J. Kerby Anderson. *Origin Science: A Proposal for the Creation-Evolution Controversy*. Grand Rapids, Mich.: Baker Book House, 1987.

Giddings, Luther Val. "Penetrating Muddied Waters: Creationism and Evolution." *Dialogue: A Journal of Mormon Thought* 19.1 (1986): 172–179.

Gieryn, Thomas F., George M. Bevins, and Stephen C. Zahr. "Professionalization of American Scientists: Public Science in the Creation/Evolution Trials." *American Sociological Review* 50 (1985): 392–409.

Gilkey, Langdon B. "Creationism: The Roots of the Conflict." *Science, Technology, and Human Values* 7 (1982): 67–71. Revised and reprinted in *Is God a Creationist?* ed. Roland Mushat Frye, 55–67. New York: Charles Scribner's Sons, 1983; and in *Christianity and Crisis* 42.1 (1982): 108–115.

———. "The Creationism Issue: A Theologian's View." In *Science and Creation*, ed. Robert W. Hanson, 174–188. New York: Macmillan, 1986.

———. *Creationism on Trial: Evolution and God at Little Rock*. Minneapolis: Winston Press, 1985.

———. "The Creationist Controversy: The Interrelationship of Inquiry and Belief." In *Creationism, Science, and the Law*, ed. Marcel C. La Follette, 129–137 Cambridge: MIT Press, 1983.

———. "Evolution and the Doctrine of Creation." In *Science and Religion*, ed. Ian Barbour. 159–181. New York: Harper and Row, 1968.

———. *Maker of Heaven and Earth: A Study of the Christian Doctrine of Creation*. Christian Faith Series. New York: Doubleday, 1959.

———. "Religion and Science in an Advanced Scientific Culture." *Zygon* 22 (1987): 165–178.

Gillespie, Neil C. *Charles Darwin and the Problem of Creation*. Chicago: University of Chicago Press, 1979.

Gish, Duane. "Creation, Evolution and Public Education." In *Evolution versus Creationism*, ed. J. Peter Zetterberg, 177–191. Phoenix, Ariz.: Oryx Press, 1983.

———. "Creation, Evolution, and the Historical Evidence." In *But Is It Science?* ed. Michael Ruse, 266–282. Buffalo, N.Y.: Prometheus Press, 1988.

———. *Evolution: The Challenge of the Fossil Record*. El Cajon, Calif: Creation-Life Publishers, 1985.

———. *Evolution? The Fossils Say No!* Public school ed. San Diego: Creation-Life Publishers, 1978.

———. "The Genesis War." *Science Digest* 89.9 (1981): 82–87.

———. "It Is Either 'In the Beginning, God'—or '. . . Hydrogen.'" *Christianity Today* 27.16 (1982): 28–33.

———. "A Reply to Gould." *Discover* 2.7 (1981): 6.

————. "The Scopes Trial in Reverse." *Humanist* 37.6 (1977): 50–51.

————. "What Actually Occurred at the Trial." *Impact* (Institute for Creation Research), no. 105 (n.d.).

Gish, Duane, and Isaac Asimov. "The Genesis War." *Science Digest* 89.9 (1981): 82–87.

Godfrey, Laurie R., ed. *Scientists Confront Creationism.* New York: Norton, 1983.

Gordon, Robert M. "*McLean v. Arkansas Board of Education*: Finding the Science in 'Creation Science.'" *Northwestern Law Review* 77 (1982): 374–402.

Gould, Stephen Jay. *Bully for Brontosaurus.* New York: Norton, 1991.

————. "Creationism: Genesis vs. Geology." In *Science and Creationism*, ed. Ashley Montagu, 126–135. New York: Oxford University Press, 1984.

————. *Eight Little Piggies.* New York: Norton, 1993.

————. *Ever Since Darwin.* New York: Norton, 1977.

————. "Genesis and Geology." *Natural History* 97 (1988): 12–20.

————. *Hen's Teeth and Horse's Toes.* New York: Norton, 1983.

————. "Impeaching a Self-Appointed Judge." [Review of *Darwin on Trial* by Phillip E. Johnson. *Scientific American* (1992): 118–121.

Grabiner, Judith V., and Peter D. Miller. "Effects of the Scopes Trial." *Science* 85 (1974): 832–837.

Greene, John C. "Darwin as a Social Evolutionist." *Journal of the History of Biology* 10 (1977): 1–27.

Habgood, John. "Myths of Religion, Myths of Science." *Nature* 300 (1982): 118.

Halloran, S. Michael, and Merrill D. Whitburn. "Ciceronian Rhetoric and the Rise of Science: The Plain Style Reconsidered." In *The Rhetorical Tradition and Modern Writing*, ed. James J. Murphy, 58–72. New York: Modern Language Association, 1982.

Hanson, Robert W., ed. *Science and Creation: Geological, Theological, and Educational Perspectives.* Issues in Science and Technology series. New York: Macmillan, 1986.

Harrold, Francis B., and Raymond A. Eve, eds. *Cult Archeology and Creationism: Understanding Pseudo-Scientific Beliefs about the Past.* Iowa City: University of Iowa Press, 1987.

Hauerwas, Stanley. *Christian Existence Today: Essays on Church, World, and Living in Between.* Durham, N.C.: Labyrinth Press, 1988.

Hayes, Arthur Garfield. "The Scopes Trial." In *Evolution and Religion: The Conflict between Science and Theology in Modern America*, ed. Gail Kennedy, 35–52. Boston: D. C. Heath, 1957.

Hodge, M.J.S. "England." In *The Comparative Reception of Darwinism*, ed. Thomas F. Glick, 3–31. Austin: University of Texas Press, 1974.

Hull, David L. "Charles Darwin and 19th-Century Philosophy of Science." In *Foundations of Scientific Method: The Nineteenth Century*, ed. Ronald N. Giere and Richard S. Westfall, 115–132. Bloomington: Indiana University Press, 1973.

Hunter, James Davison. *Culture Wars*. New York: Basic Books, 1991.

Hye, Allen E. "A Tennessee Morality Play: Notes on *Inherit the Wind*." *Markham Review* 9 (1979): 17–20.

Inherit the Wind. Produced and directed by Stanley Kramer. Screenplay by Nathan E. Douglas and Harold Jacob Smith. United Artists, 1960. Videocassette.

Irvine, William. *Apes, Angels, and Victorians: The Story of Darwin, Huxley, and Evolution*. New York: McGraw-Hill, 1955.

Jacobs, Alan. "The Unnatural Practices of Stanley Fish." *South Atlantic Review* 55 (1990): 87–97.

James, William. "The Will to Believe." In *Essays in Pragmatism*, 88–109. New York: Hafner, 1948.

Jeffrey, Duane E. "Dealing with Creationism." *Evolution* 37 (1983): 1097–1100.

Jensen, J. Vernon. "Return to the Wilberforce-Huxley Debate." *British Journal for the History of Science* 21 (1988): 161–179.

Johnson, Barbara. Introduction to *Dissemination*, by Jacques Derrida, vii–xxxiii. Trans. Barbara Johnson. Chicago: University of Chicago Press, 1981.

Johnson, Phillip E. *Darwin on Trial*. Washington, D.C.: Regnery Gateway, 1991.

———. "Evolution as Dogma: The Establishment of Naturalism," *First Things*, October 1990, 15–22.

———. "Liberals and Creationists." 1989. Typed ms.

Kaplan, Morris Bernard. "The Trial of John T. Scopes." In *Six Trials*, ed. Robert S. Brumbaugh, 107–119. New York: Thomas Y. Crowell, 1969.

Keith, Bill. *Scopes II: The Great Debate*. Shreveport, La.: Huntington House, 1982.

Kitcher, Philip. *Abusing Science: The Case against Creationism*. Cambridge: MIT Press, 1982.

Kuhn, Thomas S. *The Structure of Scientific Revolutions*. 2nd ed. Chicago: University of Chicago Press, 1970.

La Follette, Marcel C. "Creationism in the News: Mass Media Coverage of the Arkansas Trial." In *Creationism, Science, and the Law*, ed. Marcel C. La Follette, 189–208. Cambridge: MIT Press, 1983.

———, ed. *Creationism, Science, and the Law: The Arkansas Case*. Cambridge: MIT Press, 1983.

———, ed. *Science, Technology, and Human Values* 7.40 ("Tenth Anniversary Issue with a Special Section on Creationism, Science, and the Law") (summer 1982).

Larson, Edward J. *Trial and Error: The American Controversy over Creation and Evolution*. Updated ed. New York: Oxford University Press, 1989.

Latour, Bruno. *The Pasteurization of France*. Cambridge: Harvard University Press, 1988.

Laudan, Larry. "The Demise of the Demarcation Problem." In *But Is It Science?* ed. Michael Ruse, 337–350. Buffalo, N.Y.: Prometheus Press, 1988.

Lawrence, Jerome, and Robert E. Lee. *Inherit the Wind.* New York: Bantam, 1955.

Lazar, Ernie. *Creation/Evolution Bibliography/Directory.* Sacramento, Calif.: Author, [1987].

Leedes, Gary C. "Monkeying Around with the Establishment Clause and Bashing Creation-Science." *University of Richmond Law Review* 22 (1987): 149–186.

Levine, Lawrence W. *Defender of the Faith: William Jennings Bryan: The Last Decade, 1915–1925.* New York: Oxford University Press, 1965.

Lippmann, Walter. *American Inquisitors: A Commentary on Dayton and Chicago.* New York: Macmillan, 1928.

Livingstone, David N. *Darwin's Forgotten Defenders: The Encounter between Evangelical Theology and Evolutionary Thought.* Grand Rapids, Mich.: William B. Eerdmans, 1987.

Lubenow, Marvin L. *"From Fish to Gish": The Exciting Drama of a Decade of Creation-Evolution Debates.* San Diego: CLP Publishers, 1983.

Lucas, J. R. "Wilberforce and Huxley: A Legendary Encounter." *Historical Journal* 22 (1979): 313–330.

MacIntyre, Alasdair. *Whose Justice? Which Rationality?* Notre Dame, Ind.: University of Notre Dame Press, 1988.

Marsden, George M. "A Case of the Excluded Middle: Creation versus Evolution in America." In *Uncivil Religion: Interreligious Hostility in America,* ed. Robert N. Bellah and Frederick E. Greenspan, 132–155. New York: Crossroads Press, 1987.

———. "Creation versus Evolution: No Middle Way." *Nature* 305 (1983): 571–574.

———. *Fundamentalism and American Culture: The Shaping of Twentieth-Century Evangelicalism, 1870–1925.* New York: Oxford University Press, 1980.

———. "Secularism and the Public Square." In *The Best of "This World,"* ed. Michael A. Scully, 103–117. Lanham, Md.: University Press of America, 1986.

———. "Understanding Fundamentalist Views of Science." In *Science and Creationism,* ed. Ashley Montagu, 95–116. New York: Oxford University Press, 1984.

Masters, Edgar Lee. "The Christian Statesman." *American Mercury* 3 (1924): 385–398.

Mather, Kirtley F. "The Scopes Trial and Its Aftermath." *Journal of the Tennessee Academy of Science* 52 (1982): 2–9.

May, Robert M. "Creation, Evolution, and High School Texts." In *Science and Creationism,* ed. Ashley Montagu, 306–310. New York: Oxford University Press, 1984.

Mayer, William V. "The Emperor's New Clothes—Sold Again." *Humanist* 37.6 (1977): 52–53.

———. "Evolution: Yesterday, Today, Tomorrow." *Humanist* 37.1 (1977). 16–22.

McGowan, Chris. *In the Beginning: A Scientist Shows Why the Creationists Are Wrong*. Toronto: Macmillan of Canada, 1983.

McIntosh, Wayne V. "Litigating Scientific Creationism, or 'Scopes' II, III, ... " *Law and Policy* 7 (1985): 378–394.

McIver, Thomas. *Anti-Evolution: An Annotated Bibliography*. London: McFarland, 1988.

———. "A Creationist Walk through the Grand Canyon." *Creation/Evolution*, issue 20 (spring 1987): 1–42.

McLean v. Arkansas Board of Education. 529 F. Supp. 1255 (F.D. Ark. 1982). Transcript.

McKown, Delos B. "Creationism and the First Amendment." *Creation/Evolution*, issue 7 (winter 1982): 24–32.

McMullin, Ernan. "Introduction: Evolution and Creation." In *Evolution and Creation*, ed. Ernan McMullin, 1–55. Notre Dame, Ind.: University of Notre Dame Press, 1985.

———. "Values in Science." *Philosophy of Science Association* 2 (1982): 1–25.

———, ed. *Evolution and Creation*. Notre Dame, Ind.: University of Notre Dame Press, 1985.

Mencken, H. L. *The Impossible H. L. Mencken: A Selection of His Best Newspaper Stories*. Ed. Marion Elizabeth Rodgers. New York: Doubleday/Anchor Books, 1991.

———. "In Memoriam: W.J.B." In *Prejudices: Fifth Series*, 69–74. New York: Alfred A. Knopf, 1926.

Merkel, Jim. "Judge Who Ruled against Creation Law Speaks Out." *Christianity Today* 26.19 (1982): 48, 50, 52.

Midgley, Mary. "Evolution as a Religion: A Comparison of Prophecies." *Zygon* 22 (1987): 179–194.

———. *Evolution as a Religion: Strange Hopes and Stranger Fears*. London: Methuen, 1985. Milton, John. *Areopagitica: The Prose of John Milton*. Ed. J. Max Patrick, 265–334. New York: New York University Press, 1968.

Mitchell, W.J.K., ed. *Against Theory: Literary Studies and the New Pragmatism*. Chicago: University of Chicago Press, 1985.

Montagu, Ashley, ed. *Science and Creationism*. New York: Oxford University Press, 1984.

Moore, James R. "1859 and All That: Remaking the Story of Evolution-and-Religion." In *Charles Darwin, 1809–1882: A Centennial Commemorative*, ed. Roger G. Chapman and Cleveland T. Duval, 167–194. Wellington, New Zealand: Nova Pacifica, 1982.

———. "An Equivocal Heritage." Review of Livingstone's *Darwin's Forgotten Defenders*. *Science* 240 (1988): 1049–1050.

————. "Geologists and Interpreters of Genesis in the Nineteenth Century." In *God and Nature: Historical Essays on the Encounter between Christianity and Science*, ed. David C. Lindberg and Ronald L. Numbers 323–350. Berkeley: University of California Press, 1986.

————. "Interpreting the New Creationism." *Michigan Quarterly Review* 22 (1983): 321–334.

————. *The Post-Darwinian Controversies: A Study of the Protestant Struggle to Come to Terms with Darwin in Great Britain and America, 1870–1900.* Cambridge: Cambridge University Press, 1979.

Morris, Henry M. *History of Modern Creationism.* San Diego: Master Book Publishers, 1984.

————. *The Long War against God: The History and Impact of the Creation/Evolution Conflict.* Grand Rapids, Mich.: Baker Book House, 1989.

————, ed. *Scientific Creationism.* Public school ed. San Diego: CLP Publishers, 1974.

Mullins, Morrell E. "Creation Science and McLean v. *Arkansas Board of Education:* The Hazards of Judicial Inquiry into Legislative Purpose and Motive." *University of Arkansas at Little Rock Law Journal* 5 (1982): 345–396.

National Academy of Sciences. *Science and Creationism: A View from the National Academy of Sciences.* Washington, D.C.: National Academy Press, 1984.

Nelkin, Dorothy. *The Creation Controversy: Science or Scripture in the Schools.* New York: Norton, 1982.

————. "From Dayton to Little Rock: Creationism Evolves." In *Creationism, Science, and the Law*, ed. Marcel C. La Follette, 74–85. Cambridge: MIT Press, 1983.

————. "Science, Rationality, and the Creation/Evolution Dispute." In *Science and Creation*, ed. Robert W. Hanson, 33–45. New York: Macmillan, 1986.

————. *Science Textbook Controversies and the Politics of Equal Time.* Cambridge: MIT Press, 1977.

Newell, Norman D. *Creation and Evolution: Myth and Reality.* New York: Columbia University Press, 1982.

Numbers, Ronald L. "Creationism in 20th-Century America." *Science* 218 (1982): 538–544.

————. "The Creationists." In *God and Nature: Historical Essays on the Encounter between Christianity and Science*, ed. David C. Lindberg and Ronald L. Numbers, 391–423. Berkeley: University of California Press, 1986.

————. *The Creationists.* New York: Alfred A. Knopf, 1992.

————. "The Dilemma of Evangelical Scientists." In *Evangelicalism and Modern America*, ed. George M. Marsden, 150–160. (Grand Rapids, Mich.: William B. Eerdmans, 1984.

Ogden, C. K., and I. A. Richards. *The Meaning of Meaning.* 8th ed. New York: Harcourt Brace, 1956.

Overton, William R. Decision. *McLean v. Arkansas Board of Education*. 529 F. Supp. 1255 (E.D. Ark. 1982).

———. "Speech to Pennsylvania Appellate Judges." Bucknell University, Lewisburg, Pa., July 29, 1982. Typed ms. Law Clerk's File, *McLean v. Arkansas Board of Education*, 529 F. Supp. 1255 (E.D. Ark. 1982).

Page, Ann L., and Donald A. Clelland. "The Kanawha County Textbook Controversy: A Study of the Politics of Lifestyle Concern." *Social Forces* 57 (1978): 265–281.

Parker, Franklin. "Behind the Creation-Evolution Controversy." *College Board Review*, no. 123 (spring 1982): 18–21.

Placher, William C. *Unapologetic Theology*. Louisville: Westminster/John Knox, 1989.

Provine, William B. "Response to Phillip Johnson." *First Things* (October 1990):23–24.

Ridley, Mark. *The Essential Darwin*. London: Allen and Unwin, 1987.

Rorty, Richard. "Science as Solidarity." In *The Rhetoric of the Human Sciences: Language and Argumentation in Scholarship and Public Affairs*, ed. John S. Nelson, Alan Megill, and Donald N. McCloskey, 38–52. Madison: University of Wisconsin Press, 1987.

Rosenthal, Peggy. *Words and Values: Some Leading Words and Where They Lead Us*. New York: Oxford University Press, 1984.

Ruse, Michael. "Critical Notice: Philip Kitcher, *Abusing Science: The Case Against Creationism*." *Philosophy of Science* 51 (1984): 348–354.

———. *The Darwinian Revolution: Science Red in Tooth and Claw*. Chicago: University of Chicago Press, 1979.

———. *Darwinism Defended: A Guide to the Evolution Controversies*. Reading, Mass.: Addison-Wesley, 1982.

———. "A Philosopher at the Monkey Trial." *New Scientist* 93 (1982): 317–319.

———. "A Philosopher's Day in Court." In *Science and Creationism*, ed. Ashley Montagu, 311–342. New York: Oxford University Press, 1984. Also published in *But Is It Science?* ed. Michael Ruse, 13–35. Buffalo, N.Y.: Prometheus Press, 1988.

———. *Taking Darwin Seriously: A Naturalistic Approach to Philosophy*. Oxford, U.K.: Basil Blackwell, 1986.

———, ed. *But Is It Science? The Philosophical Question in the Creation/Evolution Controversy*. Buffalo, N.Y.: Prometheus Press, 1988.

Sandeen, Ernest. *The Roots of Fundamentalism: British and American Millenarianism, 1800–1930*. Grand Rapids, Mich.: Baker Book House, 1970.

Saussure, Ferdinand de. *Course in General Linguistics*. Ed. Charles Bally and Albert Sechehaye. Trans. Wade Baskin. New York: McGraw-Hill, 1959.

Shipley, Maynard. *The War on Modern Science: A Short History of Fundamentalist Attacks on Evolution and Modernism*. New York: Alfred A. Knopf, 1927.

Shumach, Murray. "'Monkey Trial' Staged." *New York Times*, 21 April 1955, II: 3.

Skinner, B. F. *Verbal Behavior.* New York: Appleton-Century-Crofts, 1957.

Smith, Barbara Herrnstein. *Contingencies of Value: Alternative Perspectives for Critical Theory.* Cambridge: Harvard University Press, 1988.

———. *On the Margins of Discourse: The Relation of Literature to Language.* Chicago: University of Chicago Press, 1978.

Smith, Willard H. "William Jennings Bryan at Dayton: A View Fifty Years Later." *Proceedings of the Indiana Academy of the Social Sciences,* 3rd ser. 10 (1975): 80–88.

Smout Kary D. "Attacking (Southern) Creationists: In *Religion in the Contemporary South: Diversity, Community, and Identity,* ed. O. Kendall White, Jr., and Daryl White, 59–66. Athens: University of Georgia Press, 1995.

Strossen, Nadine. "'Secular Humanism' and 'Scientific Creationism': Proposed Standards for Reviewing Curricular Decisions Affecting Students' Religious Freedom." *Ohio State Bar Journal* 47 (1986): 333–407.

Sullivan, LeRoy L. "The Arkansas Landmark Court Challenge of Creation Science." *College Board Review,* no. 123 (spring 1982): 12–17, 32–35.

Szasz, Ferenc Morton. *The Divided Mind of Protestant America, 1880–1930.* University: University of Alabama Press, 1982.

———. "The Scopes Trial in Perspective." *Tennessee Historical Quarterly* 30 (1971): 288–298.

Titone, Vito J. "Only Fools and Dead People Never Change Their Opinions." *Buffalo Law Review* 26 (1987): 193–209.

Unger, Robert Mangabeira. *Knowledge and Politics.* New York: Macmillan/Free Press, 1975, 1984.

Villarreal, Judith A. "God and Darwin in the Classroom: The Creation/Evolution Controversy." *Chicago-Kent Law Review* 64 (1988): 335–374.

Warren, Robert Penn. *All the King's Men.* New York: Harvest/HBJ, 1946.

Webb, George W. Review of La Follette's *Creationism, Science, and the Law* and Nelkin's *The Creation Controversy. Isis* 75 (1984): 580–581.

Weinberg, Stan, ed. *Reviews of Thirty-One Creationist Books.* Syosset, N.Y.: National Center for Science Education, 1984.

Weinshank, Donald J., Stephan J. Ozminski, Paul Ruhlen, and Wilma M. Barrett, eds. *A Concordance to Charles Darwin's Notebooks, 1836–1844.* Ithaca, N.Y.: Cornell University Press, 1990.

Whitcomb, John C., and Henry M. Morris. *The Genesis Flood: The Biblical Record and Its Scientific Implications.* Philadelphia: Presbyterian and Reformed Publishing, 1961.

Whitehead, John W., and John Conlan. "The Establishment of the Religion of Secular Humanism and Its First Amendment Implications." *Texas Tech Law Review* 10 (1978): 1–66.

Whitney, William Dwight. *Language and the Study of Language.* 5th ed. New York: Charles Scribner's Sons, 1887.

[Wilberforce, Samuel.] Review of *On the Origin of Species*. *Quarterly Review* 108 (1860): 225–264.

Williams, Raymond. *Keywords: A Vocabulary of Culture and Society*. Rev. ed. New York: Oxford University Press, 1983.

Wills, Garry. *Under God: Religion in American Politics*. New York: Simon and Schuster, 1990.

Wilson, David B., ed. *Did the Devil Make Darwin Do It? Modern Perspectives on the Creation-Evolution Controversy*. Ames: Iowa State University Press, 1983.

Wilson, John F. "Original Intent and the Church-State Problem." Lecture, Duke University Divinity School, April 6, 1989.

The World's Most Famous Court Trial. Unedited transcript of the Scopes Trial. Cincinnati: National Book Company, 1925.

Young, Robert M. "Darwin's Metaphor: Does Nature Select?" *Monist* 55 (1971): 442–503.

———. "Evolutionary Biology and Ideology: Then and Now." *Science Studies* 1 (1971): 177–206.

Zetterberg, J. Peter, ed. *Evolution versus Creationism: The Public Education Controversy*. Phoenix, Ariz. Oryx Press, 1983.

Index

About the Author

KARY DOYLE SMOUT is Associate Professor of English at Washington and Lee University. Among his earlier publications are contributions to *American Speech*, *Legal Writing*, and *Composition Studies*.

ISBN 0-275-96262-8

HARDCOVER BAR CODE